Dairy and Farm Animal Breeding

Dairy and Farm Animal Breeding

David Thompson
Editor

KOROS PRESS LIMITED
London, UK

Dairy and Farm Animal Breeding

© 2012

Printed in 2017 for Sale in the Indian Subcontinent

Published by
Koros Press Limited
3 The Pines, Rubery B45 9FF, Rednal,
Birmingham, United Kingdom

Tel.: +44-7826-930152
Email: info@korospress.com
www.korospress.com

ISBN: 978-1-78163-015-0

Editor: David Thompson

Printed in U.K.

British Library Cataloguing in Publication Data
A CIP record for this book is available from the British Library

10 9 8 7 6 5 4 3 2 1

Exclusively distributed by CBS Publishers & Distributors Pvt. Ltd.

Sales & Distribution Rights only for India, Pakistan, Bangladesh, Sri Lanka, Nepal and Bhutan.This book is not to be sold outside these territories.

Contents

(vi)

Preface

Objectives in the breeding of farm animals, role of animal breeder in improving livestock economy. Existing livestock genetic resources with particular emphasis on production potentials; comparison with foreign breeds and basis for the superior/inferior performance. Breeding policies for cattle, buffaloes, sheep and goats and other livestock species in Pakistan. Strategies for improved use and conservation of livestock. Inbreeding and crossbreeding for livestock improvement: Development and uses of inbred lines in poultry. Heterosis: genetic aspects, exploitation, crossbreeding systems. Detection and elimination of genetic defects and lethals from the stock. Maintenance of performance records, their standardization and optimal use. Large scale genetic evaluation. Interpreting genetic evaluation information. Exercises on the evaluation of productive traits of various breeds of livestock. Exercises on standardization of records, genetic evaluation and selection, Computer usage for animal evaluation. Review paper required. Visits to different breeding farms.

Genetic basis of variation in reproductive functions. Relationship of reproduction to genetic improvement in domestic animals; Pre-and post natal selection of genetic and reproductive material; the problem of reduced fertility and reproductive efficiency and its relation to lifetime productive efficiency; Measures of reproductive efficiency in farm animals; Genetic and environmental basis of the differences in reproductive efficiency; Fertility level of males for use in artificial insemination; Reduction in estrous period, quiescent heat and anestrus; Embryonic losses; Issues related to prolificacy; Manipulation of reproductive cycle, Feeding and management practices in improving fertility traits; Augmenting reproductive rate, Application of biotechnology for improving reproductive efficiency. Exploitation of genetic variation in reproductive performance for improvement. Practice on the slaughter house material and experimental live animals. Rectal palpation of genital organs at various stages of reproductive cycle. Practice in various methods of heat detection; correction of reproductive problems. Hormonal application and other techniques.

Practice in pregnancy diagnosis in various livestock species. Measurement of reproductive fitness in males. Determining reproductive efficiency in males and females.

Statistical concepts in animal genetics. Review of matrix algebra and linear models, elementary probability and its application to Mendelian inheritance. Metric characters, population mean, average effect, breeding value, dominance deviations, interaction deviations. Sources of genetic variation in single and multilocus traits, sources of environmental variation. Estimation of variance components: parent-offspring regression, sib analysis, cross-classified designs, genotype x environment interaction, maternal effects and other genetic models. Estimation of breeding values.

This book is offered with the hope that it will be a constant companion for students of this subject.

—*Editor*

1

Dairy Cattle Breeding

The Division is involved in conducting research in the areas of Animal Genetics & Breeding, Livestock Production and Management and Biotechnology related to Molecular Genetics. The main thrust areas in research include genetic improvement of crossbred and Sahiwal cattle and Murrah buffaloes by progeny testing of breeding males and judicious selection of females, faster multiplication of superior germplasm, studies on selection criteria, development of optimum breeding plan and genetic polymorphism studies at molecular level.

Development of high-level manpower in the field of Animal Genetics and Breeding constitutes another important mandate of the Division. The Division provides assistance to KVK/TTC Training Programmes, consultancy services to farmers and various organizations, supplies superior germplasm in the form of frozen semen and breeding males to farmers, developmental agencies and state governments.

Centre for Advanced Studies (Animal Genetics & Breeding)

The Centre for Advanced Studies (CAS) in Animal Genetics and Breeding established at Division of Dairy Cattle Breeding during eighth plan continued its activities on strengthening research, teaching and imparting training to scientists/teachers from research Institutes and state agricultural universities in advanced areas of Animal Genetics and Breeding. A total of 26 National Training Programmes have so far been organized under the aegis of CAS in the Division. The Divisional Library under CAS has also been strengthened.

Structuring of Research Labs/Cells/Sections/Units

The organizational structure for research facilities consists of

Animal Breeding Cell, Molecular Genetics Labs, Computer Cell, Livestock Record Cell.

Research Programme: Genetic improvement of milch Animals through identification and dissemination of superior germplasm by application of emerging reproductive and molecular technologies.

Research Highlights

A total of 32 Sahiwal (26.45 %), 76 KF (32 %) and 54 Murrah (38.29 %) females were identified as elite animals. The best lactation milk yield of elite Sahiwal, Karan Fries and Murrah buffaloes ranged between 2305 – 4725 kg, 5018-8338 kg and 2530 to 3828 kg, respectively.

The VI-Set of 6 Sahiwal bulls was evaluated on the basis of performance of their daughters using contemporary comparison method. Two bulls were selected as proven bulls with their Expected Breeding Values of 1652 and 1642 kg, respectively.

Genetic evaluation of VI-Set of Murrah bulls using contemporary comparision method was done and one bull having sire index of 1972 kg and dam's best lactation milk yield of 3512 kg was declared proven.

Artificial neural network (ANN) was trained to predict lifetime milk yield in Sahiwal cattle on the basis of five first lactation traits. Higher estimates of accuracy of prediction of lifetime milk yield from artificial neural networks in comparison to multiple regression analysis revealed that ANN methodology can be used as an alternative approach to predict lifetime milk production in Sahiwal cattle. A total of 2494 and 1942 A.I were performed in KF and Murrah buffaloes resulting in 48.9% and 45.80 % conception rate under field conditions. A total of 875 KF calves (455 males and 420 females) and 743 Murrah calves (403 males and 340 females) were born in the farmers' herds.

The test day monthly milk yields (TDMMY) of Karan Fries (KF) cows maintained up to third lactation at NDRI were used to fit different lactation curves. Wilmink function revealed lower initial TDMMY in first lactation of KF cows. The R2 value of Wood and Wilmink function ranged from 80 to 91 % across three lactations.

Prediction of first lactation 305-day milk yield based on test day milk yield (TDYs) in Murrah buffaloes revealed that the accuracy of prediction varied between 42% (TDY 1) – 67%(TDY 6). The regression equation with three variables TDY 3, TDY 6 and TDY 9 was considered most appropriate for prediction of first lactation 305-day milk yield with 91% accuracy.

The loci of DRB3, CXCR2 and beta-defensin were highly polymorphic in Sahiwal cattle and Murrah buffaloes.

PCR-RFLP analysis of exon V, VI and intron V of Pit 1 gene in Sahiwal cattle revealed AA, AB, and BB genotypes with frequencies of 0.017, 0.231and 0.752 respectively. Frequencies of A and B alleles were 0.133 and 0.867 respectively. AA genotypes showed highest milk yields followed by BB and AB types.

In Murrah buffaloes PCR-RFLP analysis of TLR 4 gene revealed AA, AB and BB genotypes using *Alu I* and *Bsp 1286I* restriction enzymes.

PCR-SSCP studies on TNP-1 and TNP-2 genes in KF, SW and MU bulls revealed three SNP's at 205, 340 and 346 bp position of intronic region between two patterns on TNP-1 gene, two SNP's at 182 bp and 186 bp position of intronic region between two patterns of first primer of TNP-2 gene.

Animal Cytogenetics and Immunogenetics

Development in animal cytogenetics and immunogenetics of farm animals. Immunoglobulins and their types: antigen-antibody interactions, Immune response, ELISA. Major histocompatibility complex; genetics of biochemical variants and their applications; Ir-genes and concepts of disease resistance including major genes; hybridoma and its significance; concept of immuno-fertility, BoLA, BuLA, TLRs, Interleukins. Chromatin structure of eukaryotes; chromosome number and morphology in farm animals banding and karyotyping; chromosomal and genetic syndromes, DNA packing in chromosomes, Z+B DNA, FISH chromosome painting and PRINS. RH Panel Mapping.

Mutation and assays of mutagenesis; sister chromatid exchanges; recombinant DNA technique and its application in animal improvement programme. In vitro preparation of somatic metaphase chromosomes; screening of chromosomal abnormalities; microphotography and karyotyping; banding procedures for comparing the chromosomal complement, FISH and PRINS. In vivo preparation of somatic metaphase chromosomes; screening of chromosomal abnormalities; microphotography and karyotyping; banding procedures for comparing the chromosomal complement.

Molecular Genetics in Animal Breeding Unit I

Basic concept: Genesis and importance of molecular techniques; Genome organization – physical and genetic map, current status of genome maps of livestock. Molecular markers and their application; RFLP, RAPD, Microsatellite/Minisatellite markers, SNP marker, DNA fingerprinting DNA sequencing, Genome sequencing, Genomic Library, Polymerase Chain Reaction (PCR), its types (PCR-RFLP, AS-PCR etc.) and applications; Transgenesis and methods of gene transfer

Quantitative Trait Loci (QTL) mapping and its application in animal breeding, Genome scan, Candidate gene approach, Genomic selection, Marker Assisted Selection-basic concepl.. Extraction and purification of genomic DNA, Gel electrophoresis, Restriction enzyme digestion of DNA and analysis, PCR, PCR-RFLP, PCR-SSCP. Bioinformatics tool for DNA sequence analysis, Design of primer, Isolation of RNA, cDNA synthesis.

Population and Quantitative Genetics in Animal Breeding

Individual versUs population. Genetic Structure of population. Factors affecting changes in gene and genotypic frequencies and their effect on genetic structure of animal populations. Approach to equilibrium under different situations: viz. single autosomal locus with two alleles, single sex-linked locus, two pairs of autosomal linked and unlinked locion Small population: random genetic drift, effective population size, pedigreed populations, regular and irregular inbreeding systems.

Quantitative genetics-gene effects, population mean and variance and its partitioning, biometric relations between relatives. Genetic and phenotypic parameters-their methods of estimation, uses, possible biases and precision. Scale effects and threshold traits. Problems relating to gene and genotypic frequencies under different conditions. Estimation of inbreeding coefficient. Estimation of effective population size. Computation of quantitative genetic effects. Estimation of variance components. Computation of heritability, repeatability, genetic, environmental and phenotypic correlations and their standard errors.

Selection Methods and Breeding Systems

Type of selection and their genetic consequences. Response to selection and its prediction and improvement of response to selection. Theoretical aspects of accuracy and efficiency of different base of selection. Prediction of breeding value using different criteria.

Combined Selection. Correlated response to selection and efficiency of indirect selection.

Selection of several traits : Evaluation of short term and long term selection experiments viz: bidirectional selection and asymmetry of response, selection plateu and limit.

Genetic aspects and consequences of various mating systems. Effects of mating systems on mean and variance. Application of various mating systems in animal improvement. Selection for general and specific combining ability. Genetic polymorphysim and its application in genetic improvement.

Estimation of breeding values from different sources of information. Prediction of direct and correlated response to different bases of selection. Computation of breeding values using different sources of information for female and male selection. Compuatation of realized heritability and genetic correlation. Selection index: Computation, Accuracy and response in component traits.

Biometrical Techniques in Animal Breeding

Review of basic concepts in statistical inference and balanced experimental designs. Nature of structure of animal breeding data and sources of variation.

Introduction to matrix algebra, types of matrices and matrix operations. Determinants and their properties, methods of finding inverse of a matrix and their application.

ANOVA, Regression and Correlations. Linear and non-linear regression. Methods of estimating regression parameters. Fitting the best regression equation. Regression on dummy variables. Fisher's discriminant function and its application, D2-Statistics in divergent analysis. Principle of least squares. Methods of analysis of unbalanced animal breeding data. Adjustment of data. Database management in animal breeding.

Matrix applications, determinant and inverse of matrices: Estimation of variance components; Least squares method for analysis of research data; Collection, compilation, coding, transformation and analysis of animal breeding data by using above biometrical techniques. Domestic Animal Diversity in India, its origin, history and utilization. Present status and flow of Animal Genetic Resources and its contribution to livelihood security. Methodology for genotypic characterization of livestock and poultry breeds through systematic surveys. Management of breed; physical, biochemical and performance traits and uniqueness of animals of a breed; social, cultural and economic aspects of their owners/communities rearing the breed. Concept of conservation, *In-situ* and *ex-situ* (*invivo* and *in-vitro*); models of conservation; prioritization of breeds for conservation. National and international strategies for conservation of Animal Genetic Resources.

Status, opportunities and challenges in conservation of AnGR. IPR issues pertaining to animal genetic resources/animal products or by-products. Registration of livestock breeds and protection of livestock owner's rights in India.

History of dairy cattle and buffalo breeding. Breeds of cattle and buffallo and their characterization. Inheritance of important economic traits. Recording and handeling of breeding data. Standardization of records. Computation of correction factors for the adjustment of the data. Estimation of breeding values of the cows and bulls. Sire evaluation methods using single trait and multiple traits: construction of Sire indices, Sire evaluation under animal model, sire model; and maternal grand sire model. Open nucleus breeding systems with MOET. Methods of cross breeding. Breeding for type, milk quality and production efficiency. Plans for developing new breeds of dairy cattle. History of development of important breeds of dairy cattle.

Considerations in the import of exotic germplasm for breeding cattle in the tropics. Appraisal of buffalo and cattle breeding programmes. Role of breed assocaitions in dairy improvement. Performance recording – milk recording-Estimation of economic traits – Standardization of records – Index cards – Sire evaluation – Comparison of latest methods-Computation of genetic parameters – Genetic gain – Estimation of heterosis – Culling and replacement. Introduction to linear models. Construction and assumptions of linear models. Fixed, random and mixed models.

Least squares procedure for fitting the linear models. One-way classification, one-way classification with a co-variable, two-way classification (with and without interaction).

Henderson's methods for estimation of variance components, Basic concepts of maximum likelihood method. BLUE and BLUP. Models of full rank and not of full rank. Generalized inverse of a matrix. Building of models for various types of data; fitting one-way model, one-way model with a co-variable and two-way models (with and without interaction). BLUE and BLUP estimates. Generalized inverse of a matrix. Breeds–Economic traits–Prolificacy-Breeding records and standardization.

Genetic parameters – Selection of males and females – Breeding systems. Development of new breeds. Breeding policy – Breeding research – Conservation of breeds. Culling and replacement – EADR.

Recent Advances in Animal Genetics

Eukaryotic genome: Gene families, Pseudogenes SnRNPs, Gene conversion, tandemly repeated genes, Nuclear Organiser region, mRNA splicing, Minisatellites, Microsatellites and its usage. Transprosons, RNA processing Transcuplion regulation of gene expression, selective gene amplification, post transceptional regulation. The proteasome and longevity of proteins. Transgenic animals their benefits in livestock production, somatic cell nuclear transfer, transgenic animals in biomedical research, ethical consideration of transgenic animals; gene therapy and transgenic animal production. Pharming of Pharmaccutical. Radiation hybrid panels and their usage in livestock, microdissection of chromosomes, *In-situ* hybridization, chromosome painting, meiotic crossing over, genome selection; Structure and functions of major histocompatibility complex, T Cell receptor, CD4, Toll Like Receptors and their functions.

Recent Trends in Animal Breeding

Biometrical models and their analytical techniques for animal breeding data using computer application and use of programmes in the field of animal breeding. Formulation of detailed breeding plans, ongoing breed improvement programmes and their impact analysis in various species of livestock under different situations. Advanced techniques in genetic manipulation for multiplication and improvement of livestock species.

Advances in Biometrical Genetics

Analysis of longitudinal data, fixed ad random regression models. Regression on dummy variable. Classificatory problems; discriminant function, D2 analysis; principal component analysis. Use of genetic parameters for prediction of recombinant inbred lines; advances in studies of genotype environment interaction and selection indices. Generation matrix and its use in population genetics; gene mapping of QTL (quantitative trait loci).

Advances in Selection Methodology

Fundamental theorem of natural selection; Selection in finite populations effect on genetic structure and variance. Optimum designs for the estimation of genetic parameters. Design of selection experiments for testing selection theory. Methods of measurement of genetic and environmental trends. Advances in selection indices Multistage, Restricted and retrospective selection indices.

Multi-information, Empirical evaluation of selection theory: genetic slippage, limits to selection, asymmetry of response, selection experiments, effect of selection on variance. Selection for threshold traits; single and multiple trait best linear unbiased estimation (BLUE) and prediction (BLUP); selection under single and multiple trait animal models; direct and correlated response through various selection indices, relationship between BLUP and selection index; fundamentals of marker assisted selections.

Advances in Molecular Cytogenetics

Structure of eukaryotic chromosomes – Evolution of karyotype – Various in vitro cell culture techniques – Cell lines and utility – Genotoxicity.

Somatic cell genetics – Stem cell genetics – Molecular cytogenetics and gene mapping – ISH, FISH, Radiation hybrid mapping, Fibre-FISH, PRINS.

Utilisation of Non-additive Genetic Variance in Farm I

Heterosis – forms and genetic basis; detection and estimation of non-additive genetic variance – average dominance, overdominance. Partitioning of between cross variance – general combining ability, specific combining ability and reciprocal effects; methods of analysing diallel crosses; utilization of non-additive genetic variance. Crossbreeding systems – crossbreeding effects; recurrent and reciprocal recurrent selection and their forms. Development of specialized sire and dam lines; inbred lines and their maintenance; inbreeding and hybridization.

Breeding Goals

Most Danish puppies are bred in small kennel facilities housing two or three bitches each of which produces one litter per year. Compared with production animals, their selection is much less systematic and primarily performed at the level of the individual dog.

Hence, total-merit indices comprising all traits of interest are not available. Instead, to a large extent, it is up to the private breeder to decide which traits should receive most emphasis in selection decisions. However, in all breeds dogs must fulfil certain criteria to be approved as breeding animals. These criteria typically relate to type, health status and/or utility traits. In many breeds within the Danish Kennel Club, dogs must obtain the evaluation 'Good' in an official dog show in order to be approved for breeding. The evaluation is performed by an authorized show judge and focuses on the phenotypic appearance of the dog with respect to movements, quality of fur, and how well the dog's 'type' fits the description set out in the breed standard. In addition, some of the working dogs, sheepdogs and hunting dogs are evaluated for their purpose-specific performance.

Most European kennel clubs were founded in the last few decades of the nineteenth century and are thus more than 100 years old. With the establishment of kennel clubs, and with the more careful recording of stud books, each breed became a closed population, so that crossbreds could not be included in the studbook. As a result of this the breed-specific types became more or less fixed.

Genetic Evaluation and Parameters

As mentioned above, all dog breeds conform to a breed standard which in Europe is described by the FCI. In very general terms dogs can be divided into working dogs and pet/show dogs. While pet/show dogs are bred primarily for their appearance to win conformation shows, working dogs are bred for their ability to perform specific tasks. Herding, hunting and grading are among the more traditional tasks that dogs have been trained and selected to excel in. However, in our modern society dogs are used for a wealth of different tasks — for instance, within certain areas of service to humans such as search, police & rescue, and assistance to blind people. Breeding values are not computed for most traits, and selection is instead based on phenotypic performance. However, breeding values are computed for the most important complex diseases, and specific genetic tests have been and are being developed for monogenetic diseases caused by single mutations.

Genetic Evaluations for Complex Diseases

BLUP animal models are used to generate breeding values free of known environmental effects for the most important complex diseases. Canine hip dysplasia (HD) is a common inherited trait in dogs characterized by hip laxity and disconformity that leads to hip osteoarthritis during maturity and in old age. Diagnoses are based on radiography of the hips.

Radiographs are evaluated to give a score for conformation based on the congruency between the femoral heads and acetabulae according to a 5-step scale. Because HD is a complex trait, it is impossible to judge the breeding value of a dog from its phenotype at a high level of accuracy. Therefore, the BLUP animal model is used to evaluate breeding value along the same lines as its use in commercial livestock breeding. As in commercial breeding the most important characteristic in the animal model is that the breeding value of the individual animal is calculated on the basis of phenotypic information from the dog itself and from all related individuals. Based on the registered HD data, a statistical model has been established which, in addition to revealing genetic effect, corrects for sex, age and year of radiography. The heritability is assumed to be 0.25 (estimated in German Shepherds). Breeding values are converted to a relative HD-index taking the outset of the average HD-index for a given breed. The average is set at 100. Thus using dogs with an HD-index >100 will improve the HD status of the breed, whereas using dogs with an HD-index <100 will have a negative impact on the HD status of the breed.

Herniation of the intervertebral disc is another disease in dogs for which genetic evaluations have been established. In hypochondroplastic breeds, the predisposition to intervertebral disc herniation is caused by an early degenerative process which can result in disc calcification. A continuous spectrum of degenerate changes is seen both within and between breeds, suggesting a multifactorial aetiology involving the cumulative effects of multiple genes and environmental factors. The disease most commonly affects Dachshunds. Severe disc degeneration with calcification has previously been shown to be highly heritable in this breed, with heritability estimates of 0.47–0.87. The number of calcified discs at two years of age is found to be a good indicator of the severity of disc degeneration and thus may function as a measure of the risk of intervertebral disc herniation .

A breeding programme, based on the association between disc calcification and disc herniation, has been established to limit the incidence of clinical disc herniation in the Dachshund population. All dogs born after 1 January 2006 must undergo a radiographic evaluation before being used for breeding. Their breeding value is calculated using BLUP animal model, assuming heritability of 0.5 and including sex, hair-variant and year of evaluation as fixed explanatory effects. As with HD, breeding values are converted to a relative index. Only dogs with an index above 100 (the average for the breed) can be used for breeding.

Organization and Breeding Programmes

In many countries breeding programmes have been established to minimize the prevalence of specific diseases. In what follows, examples of breeding programmes established by the Danish Kennel Club for some of the Danish dog breeds are described. The examples focus on breeding for improved health and inbreeding control.

Controlling Inbreeding

The closed studbooks, and the many numerically small breeds, make inbreeding control very important, but such control is made more difficult by the relatively flat breeding structure for dogs as compared with some production animals. That is, private breeders play a significant role in mating, and especially in selection decisions. Many breeds are therefore in a situation where they have low genetic diversity and clear signs of inbreeding depression.

To address this, the Danish Kennel Club has formulated a set of 'ethical recommendations' to breeders, and parts of this document have been adapted by the breeding committee of the FCI. According to the recommendations, inbreeding coefficients of up to 6.25%, equivalent to first cousin matings, are accepted.

Intensive use of popular sires is another important factor affecting levels of inbreeding in dog populations. In some breeds a male show winner or champion can be siring the vast majority of all litters for several years leading to a serious decrease in the effective population size and an increase in the risk of inbreeding over subsequent generations. The Danish Kennel Club's ethical recommendations addresses this issue by stating that no dog should sire more than 25% of the average number of puppies born per year over the course of his entire life. In other words, if a breed registers 200 puppies per year, a dog can sire 50 puppies throughout his entire breeding career. A letter is send to the owner of a male with critically many offspring, and the owner risks being expelled from the kennel club if the male continues to be used.

Breeding Programmes for Monogenic and Complex Diseases

The second highest number of diseases with a genetic basis has been described in dogs. (The highest is in humans.) A total of 507 diseases are listed in Online Medelian Inheritance in Animals. Of these, 157 are caused by a single locus. Progressive retinal atrophy (PRA)—an inherited eye disease leading to blindness—is a good example of a monogenic disease for which mandatory phenotypic screening programmes have been established in many different breeds. PRA is inherited as an autosomal recessive trait, and the phenotypic screening programme, where clinical diagnosis is made by ophthalmoscopia, cannot identify carriers.

However, the molecular basis of PRA has now been identified in many breeds, and therefore carriers can be identified by DNA testing. Presently, the molecular basis of 86 different monogenic diseases is known; the list continues to grow steadily, leading to new opportunities to include DNA diagnostics in breeding programmes. In some breeds potential breeding dogs must be genotyped for specific monogenic genetic diseases, and it is then only dogs that either do not carry the disease allele (one copy) or do not have it in two copies that are used for breeding. The breeds and diseases with mandatory genotyping prior to breeding are listed. It is important to take population size and character into account when advising breeders/breed clubs about how to include a specific DNA diagnostic test in breeding programmes. If the test is going to be performed in a large breed with few affected and carrier individuals, both these genotypes can be excluded from breeding. However, if the test is going to be performed in a small population with a relatively large proportion of affected individuals and carriers, it is advisable to include carriers for breeding for a period of time.

Breeding values are used as selection criteria for some complex disease traits, as described in the previous section. For those breeds and diseases where breeding values are available, it is typically required that breeding animals are better than the population average.

Total-merit indexes are not available, but more than one trait is often considered in selection decisions. For instance, some working dog breeds cannot become conformation champions without having passed adequate breed-specific tests measuring their working abilities. On the other hand, a Border Collie that is a conformation champion in Australia will not necessarily be a good sheepdog, and a Border

Collie that becomes a champion at sheepdog trails might not succeed in show rings, because it has nonstandard appearance. Both in respect of conformation shows and trails established to evaluate a specific ability, dogs are scored solely on the basis of an individual phenotypic evaluation.

As a side-effect of the stringent selection within the individual dog breeds and the unfavourable genetic correlations between certain type and health traits, many breeds display a high prevalence of some diseases — including certain cancers, blindness, heart disease, cataract, epilepsy, hip dysplasia and some allergies. Some of these diseases are caused by mutations in single genes, while others are complex traits influenced by both genetic factors and the environment.

Examples of Genetic Trends

It is evident that both genetic resistance to HD and calcification have improved over the past decade: the breeding restrictions that have been imposed seem to have had a positive effect.

Future Perspectives

In recent years the domestic dog has attracted considerable attention as a resource through which the genetics of disease susceptibility, morphology and behaviour can be investigated. This is partly due to the fact that many of the diseases seen in dog populations are analogous to human diseases, and partly due to the unique population structure in dogs. Because dogs show remarkable interbreed homogeneity, coupled with striking interbreed heterogeneity, the dog offers unique opportunities to understand the genetic underpinnings of natural variation in mammals. Genetic studies of dogs are theoretically simpler and more straightforward than those conducted in complex populations, offering many of the statistical conveniences of studies performed in isolated human populations, such as those carried out in Iceland. A disadvantage, as compared with production animals, is, however, the difficulty of accounting for non-genetic factors such as different care regimes, nutrition, and so forth, provided by dog owners, although further advantages are offered by the architecture of the dog genome itself: the dog is known to have longer stretches of linkage disequilibrium, reducing the overall number of markers needed to investigate the whole genome. Taken together, these features suggest that a wealth of new genetic information will be generated in dogs in the years to come. If it is used wisely, this information will greatly benefit dog breeders. However, it may also set up more challenges for those involved in genetic counselling.

Cattle Breeding – The Success of Science

Cattle breeding involve selection of stock; insemination (both natural and artificial) as well as embryo transfers for producing genetically improved offsprings. Breeding, especially the artificial form of breeding being practiced these days has realized a long cherished dream of the scientists and it has now become possible to get the genetically superior animals to produce more offsprings than it would be possible through natural mating.

Selective Breeding of Cattle

Robert Bakewell, an 18th-century English farmer, is known to have effectively put selective breeding to use for the first time and it is to him that we owe our knowledge that for a better breed (offspring/progeny), we need to breed two good breeds (parents/ antecedents). He continued pure-breeding/inbreeding beef cattle for several generations and developed better-quality beef cattle.

Inbreeding remained the prominent form of cattle breeding till the 1950s when the positive effects of crossbreeding of cattle first came to light. Reports based on studies on crossbreeding proved that the mating of two unrelated breeds resulted in the development of a calf with HETEROSIS or Hybrid Vigour. Studies also indicated that the offspring thus resulting surpassed its parents in every respect – in health, in its growth-rate as well as in its ability to produce offsprings. There is no element of doubt that crossbreeding has proved to be a boon for beef cattle breeding. In contrast, it has so far not shown any promises for dairy animal breeding.

Other selective breeding methods being practiced today involve:

- Hybrid – first generation cross between two animals that belong to different breeds
- Composite – two hybrids of same breed-combination bred back to each other for generations.

Artificial Insemination

Artificial Insemination was that revolutionary find that wove success-stories around dairy farms. AI, wherein semen from a superior-quality bull was injected in the reproductive tract of a cow to make her pregnant proved immensely beneficial and boosted dairy production manifold times.

Initially, semen collected could be stored for just four days; by the 1950s, however, scientists evolved an innovative way for storing semen for indefinite periods – frozen semen was preserved in liquid nitrogen.

Embryo Transfer Technique

The development of the Embryo Transfer Technique is another milestone event in cattle breeding. With the arrival of this technique, it has now become possible to pass on the genetic traits of top quality female animals to future generations.

ETT involves the removal of a fertilized egg from the reproductive tract of a special breed cow into the reproductive tract of a second cow. This ensures that the top characteristics of both a superior-quality bull and superior-quality cow are inherited by the offspring calf.

Advancement in technology has made it possible to freeze and store fertilized eggs for later use. A research on an 8-day-old fertilized egg that was split to produce four genetically identical calves has bolstered the belief that some day it will be possible to impregnate countless number of cows by splitting fertilized eggs from a superior breed female cattle. The day scientists are able to achieve that feat; cattle breeding will reach its zenith.

A Computational Approach to Animal Breeding

Controlled breeding programs are common in many contexts, ranging from conservation biology and zoo animal management to agricultural livestock management (CSIRO; Schneider, 1997), to laboratory animal breeding for research purposes. In most of these contexts heuristic mating strategies are used. These heuristics are generally based on personal experience and have not been extensively evaluated or analysed. In this paper, we propose a discrete optimization model of the controlled breeding problem which allows us to mathematically evaluate the effectiveness of different breeding strategies. We use this model to analyse common mating heuristics for two specific goals of breeding programs, which lie at two extremes: maximizing diversity vs. producing a particular genotype.

The first goal we consider is to produce all possible gene combinations. This is a simplified special case of the common problem in zoo breeding programs of maximizing the genetic diversity of the population in order to prevent random and directional genetic changes (Lacy, 1994). We give lower and upper bounds on the average number of matings—i.e., the number of matings that is, on average, necessary and sufficient–to achieve this goal. We also analyse the average number of matings for an analog of a strategy used in zoo breeding programs (Lacy, 1994) and show that the average number of matings needed can be significantly reduced by modifying it slightly.

The second goal we consider is breeding an animal with a specified phenotype or genotype. This is a common goal of an agricultural animal breeding program. We compare the simplified analogs of two breeding heuristics: 1) iteratively mating animals closest overall to the target and 2) breeding for one genetic trait at a time. Both strategies are commonly used in livestock management (e.g. CSIRO). We analyse the average worst case number of matings for both strategies and show that the latter, surprisingly, is more efficient. These two objectives are representative of the goals of breeding programs in conservation biology and agricultural livestock management, and thus show the wide range of applicability of our approach. While the population parameters may vary and can change the actual number of matings for a particular strategy, the order of magnitude of the number of matings on average and the relative competitiveness of the mating heuristics remain the same. Thus, our simple computational model of the animal breeding problem provides a novel, viable and robust approach to designing and comparing breeding strategies in captive populations. Our approach is different from the existing methodology for analysis of breeding strategies in that it does not follow a step-by-step process modelling the state of a population but rather looks at any such process as an algorithm and uses algorithmic analysis techniques to evaluate and compare various strategies.

Discrete Optimization Model

In this section, we present the theoretical framework used for our analysis. Each animal is represented by a binary string of n bits, each bit corresponding to a genetic marker or a trait. The results in this paper are easily extendible to an alphabet of arbitrary size, allowing us to represent alleles, nucleotides, etc. Below we list the simplifying assumptions made in our breeding model. We discuss the impact of relaxing these assumptions in Section 4.

- Any two animals can mate, i.e., there is no gender. We can easily remove this assumption without changing our results significantly.

- A mating of 2 animals is a single atomic operation. This assumption can be relaxed in many ways, from fixing the number of matings per animal to the complicated concept of a generation.

- Each mating produces exactly one offspring. Again, our results are easy to generalize to multiple offspring, or to including the probability of success (fertility) of a mating.

- There are no deaths in the course of the breeding program.

- We use a simple Mendelian model of inheritance. We assume the aid probabilistic model on the genes (or the set of traits). That is, the outcome of a mating for each bit depends only on the values of that bit in the two parents. If a bit is the same in both parents then it is the same in the offspring. If the parents' bit values are different then the offspring has a 0 or 1 for that bit, taking its value from one parent or the other, with equal probability. We make this assumption since in most cases very little is known about the dependencies between genetic loci and the probabilities of particular outcomes (Falconer, 1981).

Given an initial population of k animals, we concentrate on two particular goals. We want to minimize the average number of matings required to achieve these goals.

Past Work

There is an extensive toolkit of methods for analysis of various aspects of breeding populations, from kinship and inbreeding coefficient estimates to calculations of loss of the genetic diversity provides a good survey of the existing methodology). Presently, analysis and comparison are often done using stochastic modelling, that models the state of the population at each time step using the variety of the genetic analysis methods mentioned above to update the state. However, in many cases, the empirical statistical information underlying the model (how much does inbreeding affect survival rates, for instance?) is very hard to obtain. It is often collected for a different animal population (e.g. the statistics of the black rhinoceros in Kenya used for populations in other parts of Africa); or the population is too small and has large variance; or the environmental conditions have changed since the information was last collected. For these and many other biological and mathematical reasons, a time-step-based stochastic approach is either not robust or is simply impossible.

Often the information sought from the modelling process is qualitative rather than quantitative. In the case of designing and evaluating strategies for captive animal breeding, the main question we want to answer is: "Which strategy is better?". This is inherently a question of algorithm analysis. In this paper, we answer this question by formulating the controlled breeding as a discrete optimization problem. For this reason, we do not make use of any intermediate stochastic models.

Breeding for Maximum Diversity

In this section, we focus on the goal of producing all possible gene combinations. We assume that the initial population can actually produce all possible binary strings. That is, we ignore the bits that appear with a single value within the population. If the value of some bit is the same in all the strings of the initial population then there is, of course, no way to produce a string with the other value for that bit (in the absence of mutation).

Thus we assume that all n bits appear with both values 0 and 1 in the initial population, so that all 2n possible strings can be created from the initial population. We now consider lower and upper bounds on the average number of matings to create all 2n possible strings. A lower bound is an absolutely necessary average number of matings, no matter what particular strategy is used.

An upper bound is a sufficient average number of matings. Commonly, but not always, an upper bound is an analysis of a particular strategy. (Clearly, if there exists a strategy that achieves the desired goal within some average number of matings then that number is sufficient and serves as an upper bound.) Ideally, if a particular strategy has an average number of matings equal to the lower bound then this strategy is optimal, since it uses no more matings, on average, than are absolutely necessary. In practice, it is often hard to prove a good lower bound. Thus, a strategy may be optimal even if its average number of matings is greater than the best known lower bound.

If the size of the initial population is k, an obvious and extremely optimistic lower bound for the maximum diversity goal is 2n–k. This bound assumes that every mating produces a new string; however, this is clearly unrealistic, since many matings produce strings already present in the population. We can show a slightly improved lower bound of $2n - k + O(n)$, where $O(n)$ is some linear function of n. For our upper bound, we propose a Greedy strategy that for each successive mating chooses the two animals with the highest probability of producing a new string. This strategy is in the spirit of that used in zoo breeding programs where the two "least inbred" animals are selected for each mating. However, we show that the Greedy algorithm has at most $2.3125(2n-k)$ matings on average, while the one used by zoos has about $n(2n - k)$— thus Greedy is more efficient by a factor of n, the number of traits. Although the Greedy strategy requires at most about 2.3 times as many matings as our lower bound, we conjecture that neither the upper and the lower bounds are tight and that the Greedy approach is actually optimal.

Upper Bound

We now present an upper bound for the maximum diversity breeding problem and compare various strategies. First we show that we can ensure that the probability of producing a new animal is always at least 1/4. To do this, we introduce the notion of binary hypercubes. An n-dimensional binary hypercube is a finite n-dimensional binary lattice that consists of nodes that correspond to binary strings of n bits and edges between them. We say that two strings are a distance t apart if they differ in t bits. The nodes in a binary hypercube are connected by an edge if they are distance 1 apart. The 3-dimensional hypercube. Notice that this hypercube has six 2-dimensional hypercubes as subcubes; for example, we say that the nodes 011,010,001,000 are spanned by the initial population 010, 001, in the sense that this subcube is the set of all strings that can be generated by that initial population. As an extreme case, each edge is a 1-dimensional subcube, spanned by the strings at its two endpoints.

We will now show that unless all the strings have been produced, there always exists a pair with probability at least 1/4 of producing a new string. Assume, to the contrary, that for every two animals in the population the probability of producing a new string is less than 1/4. This means that the hypercube spanned by any pair of animals in the population has more than 3/4 nodes already present in the population. That is, for any two present nodes distance t apart, the hypercube spanned by the two nodes has at most 2t–2 – 1 nodes not yet present in the population. Specifically, for any two present nodes distance 2 apart the entire 2-dimensional subcube they span must be present in the population. The presence of any node adjacent to this 2-dimensional subcube necessarily means that the entire 3-dimensional subcube is spanned. This follows from the fact that both 2-dimensional faces formed by the adjacent node and the 2-dimensional subcube must be entirely present and the nodes of these faces span the rest of the 3-dimensional cube's 2-dimensional faces.

Upper Bound

We now present an upper bound for the maximum diversity breeding problem and compare various strategies. First we show that we can ensure that the probability of producing a new animal is always at least 1/4. To do this, we introduce the notion of binary hypercubes. An n-dimensional binary hypercube is a finite n-dimensional binary lattice that consists of nodes that correspond to binary strings of n bits and edges between them. We say that two strings are a distance t apart if they differ in t bits. The nodes in a binary hypercube are connected by an edge if they are distance 1 apart. The 3-dimensional hypercube. Notice that this hypercube has six 2-dimensional hypercubes as subcubes; for example, we say that the nodes 011,010,001,000 are spanned by the initial population 010, 001, in the sense that this subcube is the set of all strings that can be generated by that initial population. As an extreme case, each edge is a 1-dimensional subcube, spanned by the strings at its two endpoints.

We will now show that unless all the strings have been produced, there always exists a pair with probability at least 1/4 of producing a new string. Assume, to the contrary, that for every two animals in the population the probability of producing a new string is less than 1/4. This means that the hypercube spanned by any pair of animals in the population has more than 3/4 nodes already present in the population. That is, for any two present nodes distance t apart, the hypercube spanned by the two nodes has at most 2t–2 – 1 nodes not yet present in the population. Specifically, for any two present nodes distance 2 apart the entire 2-dimensional subcube they span must be present in the population. The presence of any node adjacent to this 2-dimensional subcube necessarily means that the entire 3-dimensional subcube is spanned. This follows from the fact that both 2-dimensional faces formed by the adjacent node and the 2-dimensional subcube must be entirely present and the nodes of these faces span the rest of the 3-dimensional cube's 2-dimensional faces.

The above analysis implies that the Greedy strategy that always mates with the highest probability to produce a new animal takes at most 4(2n–k) matings on average. However, we will show that the Greedy strategy performs better than this, bringing the factor 4 down considerably.

Specifically, we now show that 2.3125 · 2n average number of matings is sufficient to produce all possible strings and that the Greedy strategy requires no more than that on average. This upper bound relies on the existence of two complementary strings in the population, i.e., strings that differ on every bit. Using the techniques of Section 3, we can obtain such a pair with at most O(n log n) matings on average, which is negligible compared to 2n. Hence, we can ignore the O(n log n) term and assume that the complementary strings indeed exist in the population already.

As we have shown earlier, there always exists a pair of animals whose probability of producing a new string is at least 1/4. Therefore, we can mate the complementary strings until there are at most 3/4 · 2n animals in the population, and then mate any pair with at least 1/4 probability of producing a new string.

Notice that the maximum probability of producing a new animal is always at least that of the above specified algorithm. Hence, the Greedy strategy has at most 2.3125 · 2n average matings. This leaves us with roughly a 2.3 factor difference between the lower and upper bounds for the average number of matings needed to produce all possible strings in the population; however, we believe that neither our lower bound nor our upper bound is tight. That is, we believe that the necessary average number of matings to produce all possible strings is greater than 2n + O(n) and that the sufficient number of matings is less than 2.3 · 2n. We believe that the lower and upper bounds meet C · 2n where C is the constant 1/(1 − 1/e) _ 1.582, and that this number of matings is achieved by the Greedy strategy on average.

Comparison with a Currently Used Strategy

The Greedy strategy always mates the two animals in the population that have the highest probability of producing an animal not yet in the population. However, the strategy currently used by the conservation breeding programs is to breed the two animals whose pedigrees differ by as much as possible; in our model, the two strings farthest apart from each other in the hypercube. We will call this the Dissimilar strategy. In particular, this strategy will breed any two complementary strings if they exist. As we discussed earlier, we can assume that two complementary strings indeed exist in the population since the number of matings needed to ensure this is negligible compared to the total number of matings needed to produce all the strings. Mating two complementary strings will create a string which is chosen uniformly at random from the set of all possible strings. We can then ask how many matings of these complementary strings are required before all strings have been created at least once. This is known as the coupon collector's problem (Cormen et al., 2001), and the answer for X = 2n different strings is X lnX = n ln 2 × 2n. This is greater than the average number of matings used by the Greedy strategy, which is a constant times 2n, roughly by a factor of n. For example, if we wish to create a population of 1, 024 genetically distinct animals with all possible combinations of n = 10 traits, the Dissimilar strategy requires about 9, 966 matings on average while the Greedy strategy requires at most 2, 312 matings on average.

The advantage of Greedy becomes even greater as n increases. Thus, Greedy provably performs significantly better than Dissimilar in terms of the average number of matings in our model. A small caveat is that Greedy requires more computational effort than Dissimilar since it must repeatedly find the pair that has the highest probability of producing a new animal at each step. However, the time it takes to do this computation is surely negligible compared to the time and effort required to actually breed two animals!

Breeding a Target Animal

We now focus on the problem of breeding an animal with a specified target set of gene variations or traits. For simplicity we assume that the target string is the all-ones string 111... 1. We also assume that this string is a possible descendant of the initial population, i.e., in the subcube spanned by the initial population; this simply means that for every i, there is some initial string whose with bit is 1. Again, we wish to find lower and upper bounds on the average number of matings required to produce this target string. In the worst case, the founder population consists of n strings, $x_1, x_2, ..., x_n$, where the i'th bit of xi is 1 and its others are 0. (If there if a initial individual y whose i'th bit is 1 and it has additional 1s in its string, then for the purposes of producing the all-ones string we can only gain by using y in lieu of xi.) In this case, an obvious lower bound on the number of matings needed to get the all-ones string is n " 1, since every xi must appear at least once as one of its ancestors.

Algorithm AddOneTrait

We now show that the above algorithm, AddOneTrait, takes $2 \cdot \log2 t$ matings on average. Without loss of generality, let x have 1s in the first t bits and y have a 1 in bit $t + 1$. After the first two matings of the original x and y we expect one of the offspring to have a 1 in the (t+1)'st bit and 1s in half of the first t bits. We choose this to be our new partner for x. In any mating of this string with x, all the common 1-bits will, of course, be retained. One of the two matings of this new string with x is average to produce a string with a 1 in the $(t + 1)$'st bit and 1s in about half of the bits which are 1 only in the string x. Thus, after two matings, we expect to have a string with a 1 in $(t + 1)$st bit and $3/4 \cdot t$ 1s in the first i bits. Continuing in this manner, after $\log2 t$ iterations on average we can produce a string with $t + 1$ 1s. Since each iteration takes two matings on average, the total average number of matings is $2 \cdot \log2 t$. We now give our algorithm for breeding a target animal, BreedTarget. We assume that the initial population consists of the xi defined above, since this is the worst case. The target string is the all-ones string, so our goal is to reach $t = n$.

Biologically, this result can be interpreted as follows. Our algorithm repeatedly adds one new trait to the population and then uses back-breeding to breed an animal with both this new trait and all the previously added traits. The downside of this strategy is that new offspring are bred back with one of the parents several times, producing a highly inbred population. However, such practices are not uncommon in agricultural animal breeding programs so we do not consider it unrealistic.

We now compare our algorithm with a greedy strategy, that always mates the pair of individuals that are currently closest to the target animal. Again, we assume the worst case founder population of n strings with a single 1 bit each. The final mating that produces the all-ones string is, on average, a mating of two strings with $n - 1$ ones each (otherwise it takes more than a constant number of matings on average to produce an all-ones string in that last step). There would need to be at least three strings with $n - 2$ ones to produce these $n-1$ ones' strings. Continuing by induction, such greedy strategy would need a total number of strings This is asymptotically more (when $n \_ 15$) than the $2n \log^2 n$ matings on average required by our algorithm.

To the best of our knowledge, we have formulated the first combinatorial computational model of the con-trolled breeding problem, allowing us to design and evaluate breeding strategies in the context of discrete optimization. We have used this model to analyse common mating heuristics for two specific goals of a breeding program: 1) breeding for maximum diversity and 2) breeding a target animal. These two goals are representative of two extreme examples of breeding programs, namely conservation biology and agricultural livestock management. We have proved upper and lower bounds for the average number of matings to achieve these breeding goals. Overall, we have demonstrated the viability and robustness of our mathematical approach to analysing controlled breeding problems.

In formulating our discrete optimization model of the breeding problem, we made many simplifying assumptions. However, most of these are easy to relax: for example, introducing the concept of gender rather than allowing any pair of animals to mate at most doubles the average number of matings for any breeding strategy. Similarly, assuming that each mating can produce up to a fixed number of offspring reduces the number of matings by no more than a constant factor. One non-trivial assumption we made is that only two animals can mate at any step, i.e., that breeding takes place serially rather than in parallel. It would be interesting to investigate various strategies that minimize the number of generations needed to achieve a breeding goal. There are several ways to define a generation in mathematical terms, each one providing a challenging optimization problem. Finally, the assumption of the iid probabilistic distribution model of genes is the most challenging from a biological point of view, since little is known about genetic transfer. As biologists learn more about this area, we can develop more realistic models of controlled breeding. Acknowledgments. We would like to thank Robert Lacy and Onnie Byers of the Conservation Breeding Specialists Group and Sarah Long from the Chicago's Lincoln Park Zoo for providing information and pointing out the relevant literature. We would also like to thank Alex

Russell, Bernard Moret, and Alla Sheffer for many useful comments. This work is supported by the National Science Foundation postdoctoral fellowship grant EIA 02-03584 (Tanya Berger-Wolf), by the National Science Foundation grant CCR-0313160 (Jared Saia), and by the Sandia University Research Program Grant No. 191445 (Jared Saia).

Different Dog Breeding Methods

Dog breeding is not as simple as you think. Many responsible pet owners favour to get purebred dogs nowadays. Even on a similar multiplies, breeding lines are inclined to produce different traits and types of dogs. As you can watch, dog breeding is a complex undertaking and without enough knowledge, you can't be an impressive dog breeder.

Breeders should be mindful of the breed standards. However, dog breeding is sometimes influenced by the personal likes or dislikes of the breeder when it comes to colour, sizes, and types. What dog breed do you favour? Does it possess its true traits? Is the visual aspect of the dog the same as what the books or other information sources say? What about its temperament? Is the breeding line of your pet healthy? You have to answer these wonders before you ultimately select a certain dog breed. Answering the calls into question is in addition vital should you determine to be concerned in the breeding business.

There are nearly all considerations in dog breeding. You ought to be prepared to invest time, effort, and money before you can learn the fruits of your labour. Try to answer the questions above and if you can't, you will need to dig in deeper. You have to investigation about the different dog breeds, the traits of different bloodlines, the visual aspect of the dam and sire, grand-dam, and grand-sire in a specific dog pedigree. Do you would like to introduce improvements in the dog breed? Do you like to retain certain traits or qualities in the dog breed?

Breeding methods vary whether you're dealing with inbreeding, out crossing, line breeding, or purebreds. You must be able to gather all the facts you have about dog breeding. Keeping educated will make you a better dog breeder. You can read books on dog breeding or you are able to do some online investigation.

Cross breeding is one of the methods of dog breeding. In this technique, dogs of different multiplies are used. You cannot expect to produce the best dog strains if you use the cross breeding method. If a complaint occurs with regards to the traits of the dogs, you cannot easily track down where the trouble originated from. The next method is out breeding. This is a complex method and only experienced breeders can perform it. If a certain dog breed has a complaint or defect, you can make improvements through out breeding. You must have a mate which posses the particular trait that you want to improve. The breed should possess such trait up to that last 3 generations. After the out breeding, the dogs are bred back to the original breeding line. New breeders aren't encouraged to do out breeding because this not a long way to begin a breeding program particularly when you have plans to produce constant multiplies with the identical soundness and temperament. Educate yourself thoroughly before you consider out breeding.

Line breeding is a different method in dog breeding. In this method, intimately linked dogs are bred to produce an identical quality traits, size, and temperaments dog strains. This technique should be performed correctly to make sure of success. Breeders who would like to concentrate mainly in one dog breed should be engaged in line breeding.

It is these that are the three dog breeding methods that you should know if you have plans of breeding dogs as a sort of hobby or even for business.

Three Types of Selective Breeding

Selective breeding is the process of purposefully breeding plants and animals for the objective of obtaining particular genetic traits. The people who engage in selective breeding are known as breeders, and the bred animals are known as breeds.

Selective Breeding Terms

When you breed plants, the results are known as cultigens, cultivars or varieties. When there is a cross of animals, the results are referred to as crossbreeds, while a cross of plants results in hybrids. Similar methods are used in animal and plant breeding. When animals with desirable traits are selected, they are bred through the process of culling for particular for traits. Culling is the process of selecting livestock based on desired criteria, and destroying the others. This is how purebreds are produced. Purebreds with a recognizable lineage are known as pedigreed, while a mix of two separate purebreds will produce crossbreeds. The three methods of selective breeding are outcrossing, inbreeding and line breeding.

Line Breeding

Line breeding is the process of breeding animals or plants that are closely related so as to "fix" or "set" desirable traits. For example, if a horse has some qualities that the breeder likes, the breeder could breed that horse with another relative so as to reinforce the desirable traits through a "pooling" of the genes. The idea is that if one animal has desirable qualities, mating it with a genetically related animal will increase those desirable traits. In human terms, linebreeding is like mating two close, but one-step-removed relatives, like first cousins, grandparent and grandchild, or uncle and niece.

Inbreeding

Inbreeding is the mating of very closely related animals in the hopes of increasing the desired traits in the next generation. Inbreeding is similar to linebreeding only that in inbreeding, the animals would be as close as parents and offspring, or siblings. Inbreeding has some serious flaws because, while it may intensify the desired traits, it will also intensify any faults in the parents. Linebreeding is a little better, because the parents are one step removed. Still, it has similar drawbacks to inbreeding, because it is still that same small gene pool that is being passed from one generation to the next.

Outcrossing

Outcrossing is the breeding of two animals or plants that are not related to each other. This means that the animals do not have any related ancestors in their pedigree for four generations or more. Outcrossing introduces new traits that are missing in the limited gene pools available in linebreeding and inbreeding. It can also "dilute" the effects of inbreeding by reducing the concentration of undesirable genes.

Better Breeding Animals with Genomic Selection

Scientists from the Faculty of Agricultural Sciences at Aarhus University have collaborated with the Pig Research Centre in making it possible to use genomic selection in Danish pig breeding.

With genomic selection the genetic potential of an animal is determined according to its DNA profile. This means that scientists can use a blood sample to get a picture of the heritable traits of an animal. With present methods it is necessary to test the various traits – and that costs time and money.

Genomic selection can make it easier to combine improvements in pig welfare with production efficiency.

Senior scientist Peer Berg, Department of Genetics and Biotechnology, points out that feed consumption can be reduced because the pigs will be bred for improved feed efficiency.

Feed consumption is a major economic factor in pig production but it is difficult to improve further using present breeding methods. DNA based technology will improve the certainty of determining the genetic potential of feed efficiency in a breeding animal.

Another example is that welfare, especially on sow farms, can be improved using DNA based technology because the genetic potential of a male or female piglet can be determined already at birth.

Today the longevity of sows is difficult to improve via breeding because longevity can only be determined in old sows, says Peer Berg.

The project is based on a newly developed chip consisting of 60,000 SNP markers. The chip's huge store of information about the animal's DNA is what opens a window on new possibilities for selecting the best animals for breeding.

Compared to the methods and technology used in breeding today, genomic selection will pave the way for big changes in Danish pig breeding.

There is a need to develop better mathematical and statistical models for calculating genomic breeding values. We also need to develop breeding plans in which genomic selection is included and which ensure the best use and implementation of the new technology, says Peer Berg.

Cloning Animals

The TA project »Potential and risks of the development and use of cloning and of genetic engineering and reproduction technology in breeding animals for research, in breeding laboratory animals and breeding productive livestock« is based on an application by the Bündnis 90/Die Grünen parliamentary group in the German Parliament (Bundestag pub. 13/7160). With the aim of improving the information basis TAB was to be commissioned to make an inventory of this complex and ethically-charged issue. The TA project "Cloning animals" was adopted in summer 1997 by the Committee for Education, Science, Research, Technology and Technology Assessment of the German

Parliament, and started at the beginning of 1998. The goal of the project was to study

- what influences the use of nucleus transfer cloning may have on biological fundamental research,
- what contributions can be expected for the various application-oriented areas of medicine,
- what impacts are evident for animal breeding and agriculture,
- and finally, which problem areas can be identified and what conclusions can be drawn.

Even during concept elaboration it became clear that restricting consideration to purely technical, medical and economic aspects of the use of cloning in animals would be unsatisfactory. Consideration of the legal and ethical aspects of cloning is also required. The report accordingly addresses the question what regulations (if any) animal cloning is subject to in Germany under current legislation and whether cloning may be subject to statutory restrictions or even prohibitions. In view of the possibilities the further development of cloning techniques may have in terms of possible human areas of application (drugs, transplants, tissue culture etc.), it also seemed necessary to consider the actions of doctors, biotechnologists and animal breeders in the context of scientific and technological advances and ethical demands. The ethical question posed was whether conventional ethical principles, classic arguments and relevant models of ethical evaluation are adequate for the moral evaluation of cloning.

The following sections review the most important results of the study for the individual issues and conclusions.

The Fundamentals: Cloning Procedures

A clone is an individual which is genetically identical with another individual. Cloning is a form of asexual reproduction which is widespread in nature. In the case of single-cell organisms and plants, it is an entirely normal process (division, vegetative reproduction), in the case of the higher vertebrates genetically identical individuals can arise naturally through the spontaneous division of embryos in the early stages of division, with the parts evolving separately into independent individuals (twins, multiple births).

There are in principle two procedures for artificial cloning of higher organisms: embryo splitting, and cloning through nucleus transplantation to egg cells or embryo cells whose genetic material has been removed (nucleus transfer). Cloning techniques come under biotechnology, specifically the (biotechnology) procedures which do not modify the genetic material in the cell nucleus. Cloning techniques are, however, generally applied not in isolation but in combination with other biotechnology and genetic engineering techniques (transgenic, which modify genetic material). Here, many biotechnological procedures are an essential element of cloning, while others are optional.

Embryo splitting and nucleus transfer differ fundamentally in terms of the technique and the degree of genetic identity achieved in the resulting embryos. Embryo splitting changes neither the age nor the (toti-)potency of the cells used. The (two) embryos from the splitting are in the same stage of development, exactly the same age as the undivided embryo would have been and genetically completely identical. The nucleus transfer technique of cloning takes a different approach by transferring the genetic program (the cell nucleus with the desired genetic material) from a totipotent blastomer or no longer totipotent cell (embryonic, foetal or even a differentiated body cell) to an unfertilised egg cell whose nucleus has previously been removed. This technique basically offers the possibility of replicating an adult individual and their genetic program. The result is a new individual whose existence does not derive from the fertilisation of an egg cell by a sperm cell.

The surprising thing of the cloning technique that resulted in Dolly in 1997 is that a mammal egg cell to which a nucleus from a differentiated body cell is transferred can develop into a complete organism. The genetic material in the cell nucleus of a differentiated body cell is functionally differentiated and modified in many ways compared to the genetic material in the cell nucleus of a fertilised egg. Previously it had been assumed that cell nuclei from differentiated or specialised body cells could not in principle be reprogrammed to develop again into an individual. The Dolly experiment has accordingly raised the problem of the need to redefine the totipotency of a cell (the developmental potential to differentiate into any type of cell and tissue).

In principle there are highly promising prospects in the application of cloning, both for biomedical fundamental research and for agriculture, particularly in combination with transgenic techniques.

However, before we reach the point where applications are relevant – and, above all, efficient – a series of important questions have to be resolved. Cloning based on nucleus transfer has not always succeeded or been successful in the long term. Many of the embryos created in this way die, not infrequently shortly before or after birth. However, the surviving animals also frequently have »deficits« which hamper their development and have a deleterious effect on their health. Today, we do not know yet in detail what the sources are for the errors which currently prevent cloning based on nucleus transfer from being efficient. The mechanisms for differential gene activation in normal reproductive and developmental processes are certainly still too little understood. The question remains what the differences are between natural development processes and those occurring after artificial nucleus transfer.

Biomedical Research and Application

Clones of higher organisms are of great interest for biomedical fundamental research and applied medical research. Currently, four possible areas of application of cloning based on nucleus transfer are basically under discussion for medical purposes.

Animals as Drug Producers

The first area is so-called »gene pharming«, i.e. the use of transgenic animals to manufacture (human) proteins with therapeutic use, e.g. in their milk. This is one of the possible main areas of application in the foreseeable future for cloning based on nucleus transfer, as this makes creating the transgenic animals more effective and specific compared with conventional techniques.

The advantages of the active ingredients from biogenetic manufacturing processes (such as insulin, blood factors or other human bodily substances) are that they can be obtained in much purer form than with conventional techniques involving animal and human intermediate products. Given the availability of such animals, production of active ingredients can be on a large scale and relatively cheap. However, there are also risks to the animals due to the genetic (transgenic) manipulation, the biological activity of the produced protein and the cloning procedure itself. Hazards to people can arise from changes in the products and possible transmission of disease (pathogens), and these have to be avoided as far as possible by careful testing of drugs.

Animal Models

Another area where cloning could be used is producing transgenic animals as animal models for human diseases. Animal models are used to study the biochemical and physiological fundamental processes, and provide valuable information for understanding these in humans, and naturally also on human diseases and possible therapies. Further, new drugs can be tested in animal models for their toxicity and pharmacological effect on humans. A major obstacle in developing animal models has been that so far we have only succeeded in the case of the mouse in integrating genetically manipulated cells into the germ track of a recipient animal so that the genetic changes can be passed on. However, the physiological and anatomical differences between mice and humans are so great that symptoms of the genetic changes induced in the mouse often do not match the symptoms observed in humans.

Cloning using nucleus transfer and somatic cells creates the possibility of inducing specific genetic changes in various species (gene targeting and gene knock-out). This would also make it possible for the first time to induce disease in transgenic large animals which could be superior to the former mouse models in terms of anatomical, physiological or genetic characteristics (depending on the disease to be studied). It is generally expected that this will contribute in the medium term to an improved understanding of the clinical picture of genetically-caused human diseases, and developing effective therapies based on this. Possibilities for research and implementation in this area should accordingly be specifically encouraged and promoted.

Breeding Endogenic Body Tissue

Cloning could also make a technical contribution to the transplantation of endogenic (autologous) tissue and in so-called cell therapy. The ideal transplant tissue is easy to identify. Its cells should be as genetically identical as possible with those of the recipient. The patient's immune system then no longer recognises them as alien, eliminating problems of rejection. One optimal solution would accordingly be creating genetically identical replacement tissue. Research findings indicate that this could be done through cloning using nucleus transfer.

There is another approach to breeding human replacement tissue which is conceivable in principle. This involves using the nucleus transfer technique to create an early embryo from which pluripotent embryonal stem cells can be obtained in vitro. In humans, however, it has not yet been possible to obtain such cells even from embryos in vitro. Such a technique would also require the ethically and legally highly questionable creation and utilisation of a human embryo, unless egg cells from animals are used as recipients for the nucleus. However, this development is still in its infancy and involves its own problems, and specifically difficult ethical questions.

Xenotransplantation

A fourth area in which the use of (transgenic) cloned animals is conceivable is xenotransplantation (transplanting animal organs into humans). However, to create »donor animals«, up to a dozen genes would have to be modified in the pig, for example. This is effectively impossible using conventional techniques of genetic modification. Cloning could now make it possible to make the desired genetic modifications to cells in vitro before cloning them using nucleus transfer to make an animal with multiple genetic modifications. Even if the »ideal« donor animal could be made in this way, however, the basic problems of rejection would probably remain. It is also not certain that the alien animal will actually perform its function in the human recipient. This also leaves the problem of animal viruses adapting to humans, with the possible consequence of epidemics (cf. TAB 1999).

Livestock Breeding and Agriculture

Biotechnological techniques have been in use for some time in animal breeding and animal production. This development dates back over 50 years to the introduction of artificial insemination. The breeding possibilities of artificial insemination quickly took on great significance (in Bavaria, for example, c. 90 % of cows and 60% of sows are artificially inseminated). Progress in breeding is, however, limited (especially in cattle) by the fact that a cow can only bear one calf a year. Embryo transfer offers the opportunity here of obtaining ten and more offspring from valuable cows in a single year. Embryo transfer is now used extensively for cattle. Other biotechnological methods, e.g. in vitro fertilisation, gender diagnosis or selection, gene diagnostics and gene transfer, are expected to lead to great advances in animal breeding if they can be applied without side effects, cost-effectively and ready for practice. Cloning itself is not a breeding technique, but one which makes possible genetically-identical replication of individuals. Cloning alone does not involve any breeding or genetic advance in the resulting clones, compared to the original individual. The decisive factors in the economic efficiency and expediency of cloning for animal breeding in agriculture are the effectiveness of the cloning technique and the (breeding) value of the genetic material available for cloning. If techniques for cloning adult animals can be developed into a routine procedure, this would also have implications for animal production, whose extent would largely be decided by the cost of cloning. As long as the procedure is still very expensive, only isolated and extremely valuable, top-performance animals will be cloned, e.g. in the event of the loss (from age or disease) of the services of a very valuable breeding animal, it could be replaced by a clone of itself.

Transgenic Clones

The expected increase in genetic knowledge of productive animals as well as the associated possibilities for creating transgenic animals combined with cloning using nucleus transfer make possible the use of new strategies in animal breeding and production. It is expected that these technologies will also make the »production« of transgenic animals with modified (agricultural) characteristics more efficient than currently possible. The most important goals for gene transfer in livestock breeding in combination with cloning are: quality enhancement, gene pharming, boosting resistance to disease, and cost reduction.

Enhanced performance through gene transfer is no longer as important nowadays in agricultural livestock, as the characteristics responsible for meat and milk production are complex and multigene, difficult to modify and also adequately handled by conventional breeding. In part, gene transfer is used in an effort to improve feed conversion or reduce fat formation, particularly in pigs. This, for example, is one aspect of the primary targets for quality improvement in animal products, as is the desired change in the composition of milk. Work is underway on increasing the protein content (specifically casein) and reducing or entirely removing lactose. This type of milk would also be tolerable for people who are lactose intolerant. Extending this approach leads to the gene pharming described above. Due to the high disease-related costs of animal factory farming, genetic modification of disease resistance in animals has great importance in animal breeding. Transferring specific disease resistance genes or deactivating sites on genes which determine specific diseases could improve animal health and hence (theoretically) the quality of animal products.

Changes to Agricultural Structures

In agricultural breeding practice the introduction of (practicable) cloning (particularly in cattle breeding) could result in a restructuring of breeding organisation. Cost considerations together with the requirements for employee qualifications would probably lead to the emergence of specialist, capital-intensive, commercially-oriented breeding companies. It is doubtful whether existing breeder associations will be able to perform the biotechnological work efficiently. Further, extensive use of cloning can be expected to lead to changes at the level of stages of production, with differing impact depending on the type and size of farm. This could reinforce the structural change in this sector and generally lead to a reduction in the number of farms and jobs in agriculture.

The possibility cannot be ruled out that the impact of cloning on the use pattern in agricultural land will reinforce the trends in the agricultural sector which have been evident since the 60s. The complete exploitation of all possible improvements in productivity by highly intensive and industrial-style operations leads to declining prices and – besides the reduction in area needed per animal – to further reduction in agricultural land while at the same time the number of livestock production farms and regions is increasing. Even in the highly competition-oriented livestock production regions, an increase in regional environmental pollution can accordingly be expected.

Legal Aspects

Among the legal aspects a particularly significant question to answer is which regulations govern animal cloning in Germany (and abroad) and under what conditions cloning is legally permitted or not permitted.

Animal Testing – Breeding for Unhealthy Traits

There is no explicit consideration of cloning technologies in the Animal Protection Act. However, cloning animals could be covered by the provisions of section 7 (1) of the Act (TierSchG), as this includes regulations on animal testing and cloning procedures are still overwhelmingly in the testing stage. The application and implications of this section are, however, debated in widely differing terms. If the removal of the nucleus from the egg is not regarded as modification of genetic material in the legal sense, the transfer of the egg to the brood animal does not constitute animal testing either. However, if one hold's the view (like e.g. the Federal Ministry of Agriculture and others) that cloning using nucleus transfer is covered by section 7 (1) sentence 2 TierSchG because this involves manipulation of genetic material and cloning tests for genetically modified animals (or brood animals) may involve pain or injury, this would make cloning tests using nucleus transfer clearly subject to approval. Cloning animals could also and specifically be restricted by section 11b TierSchG (breeding for unhealthy traits) at the point where the techniques are ready for practical application and are, for example, used in producing and breeding farm animals. This would, however, only apply where cloning was used to induce unhealthy changes in the animals which were retained in subsequent breeding.

The current legal consensus seems to be that sections 7 and 11b of the Animal Protection Act constitute statutory regulation of cloning to the extent that this could involve suffering, pain or injury for the animals. A ban on cloning would, however, only come into question if significant suffering on the part of the animals is actually demonstrable.

Violation of Constitutional Rights

Constitutionally speaking, a ban on cloning would violate the constitutional rights of researchers and professionals under Article 5 (3) (Freedom of research) and Article 12 (1) (Freedom of occupation) of the German Constitution. A ban or other limitation on cloning would also constitute interference with the constitutionally-guaranteed academic/scientific freedom. There is no evident barrier within the constitution which could justify such an intervention. Under Art. 12 (1) of the German Constitution a ban on cloning would accordingly be unconstitutional as it is irreconcilable with the public welfare and not covered by the statutory reservation of Art. 12 (1) sentence 2.

Under current conditions, cloning animals is according permissible in principle and subject only conditionally to restrictions under prevailing law.

Inclusion of Animal Protection in the Constitution

A new situation could arise if animal protection was adopted as a state goal in the Constitution. A comparison with other countries shows that constitutionally-guaranteed animal protection currently only exists in Switzerland. Animal protection acquires constitutional status there at federal level through the concept of the »dignity of creatures« (Art. 24 amendment para. 3) and sets for example a barrier to the constitutional right of freedom of research. In Germany a draft constitutional article on these lines is currently being debated. The Bundesrat (Upper House) approved on 28.11.1997 draft legislation in the 13th Bundestag (Lower House) amending the constitution by introducing animal protection as a state goal.

The Bundesrat motion aimed at adding an Article 20b to the German Constitution with the object of »respecting animals as fellow creatures and protecting them within the statutory framework against avoidable suffering and injury«. Animal protection here is mostly understood as limiting animal testing, but also with respect to factory farming, animal transport and animal slaughter. The Bundestag debated this motion without reaching a decision. A final decision is now a matter for the current (14th) legislature. If such an Article is incorporated in the Constitution, cloning animals could possibly violate a constitutionally-protected object, animal protection, as there could be an inherent constitutional barrier to Article 5 (3). It can at least be said, however, that under a constitutional guarantee of animal protection the necessary consideration of other constitutionally protected objects (such as freedom of research, academic/scientific freedom – but also freedom of occupation and guarantee of the right of property) could be achieved in an individual case of application of the law and a court ruling. Animal protection as a state goal does not rule out the use of animals by humans, but raises the requirements for the necessary justification of this use.

Ethical Aspects

The legal and legal-ethical considerations are associated with ethical considerations of fellow-creature status and an ethical judgement on animal cloning. Different positions in the social debate and assessment of animal cloning can be traced back to some extent to different fundamental values. These also determine whether animal cloning is seen as having a new quality compared with conventional animal breeding or with other new techniques in animal breeding as well. For some theologically-based positions, cloning is e.g. an interference with creation which humans have no right to undertake. Anyone who feels that animals have »intrinsic value« or »creature dignity« will generally regard cloning as morally dubious, at least. From an anthropocentric position there is above all the question of the safety of products made with the help of cloning techniques and the potential risks and hazards – ecological (reducing genetic diversity) and social (industrial mass production, concentration of capital, new dependencies). Other positions see animal cloning more as a catalyst which could further reinforce other undesired trends which are currently apparent and which result from the use of other biotechnological and genetic engineering reproductive techniques, rather than as a qualitatively new step in reproductive technology.

The different social positions represent an effort to attract support for a specific moral or ethical ideal and so possibly influence the political climate as well. Given the difficulty of reaching a moral consensus, it is necessary to consider which ethical principles a possible use of animal cloning needs to orient itself by. This means not only that a possible use of animal cloning needs to be ethically justified, but also abandoning the use of the (therapeutic) possibilities offered by this technique. In this context the focus in ethical assessment of cloning is the question whether goals of animal cloning and the means or techniques used imply interference with the sphere of interests of the animals involved, and if in this case the interference can be ethically justified in terms of the considerations described above.

Specialists in ethics generally regard the goals of biomedical research and application as paramount, with special urgency or even literally vital importance for human health, and attainable only with the help of cloning of higher animals. Goals in fundamental research can also be regarded as paramount and justifying cloning higher animals where no alternative techniques are available. However, if cloning involves considerable suffering for the animal involved, it is necessary to consider whether mere human thirst for knowledge is adequate justification, or whether justification requires specific goals, i.e. the need to avoid considerable human suffering. Goals which are regarded as subordinate in importance to the above objectives are mostly goals in livestock breeding, where these do not explicitly involve ensuring the basic human food supply.

In applied research, cloning using nucleus transfer opens up new approaches to creating transgenic animals. Some proteins with therapeutic effect can be cheaply produced in this way. Obtaining autologous replacement tissue seems particularly promising in both medical and ethical terms, and the associated research is accordingly particularly deserving of promotion. It is not clear whether it will be possible to create better test models for human diseases in livestock, but the considerable medical importance justifies increased effort and support in this area as well. Overall there appear to be a relatively high potential benefits for research and medicine from cloning using nucleus transfer.

In agriculture the (practicable) production of clones of breeding stock promises to improve animal performance and quality while simultaneously reducing production costs. It is likely that cloning techniques will add further weight to existing trends in optimising the performance potential of livestock, i.e. high-performance animals. With regard to the questions of genetic improvement and diversity in animal breeding, selection for specific performance characteristics can have the goal of standardising (breeding) livestock with the help of cloning, and hence inevitably standardising livestock generally. The resulting (desired) genetic status quo is accordingly very probably tied to a reduction in genetic diversity. Although there is in principle still great need for research on more exact documentation of the status quo, appropriate measures should be taken now to limit »artificial« production of increasing numbers of offspring of individual animals. This relates to techniques ranging from current routine artificial insemination through to cloning.

It is also likely that the introduction of cloning in combination with other reproductive and genetic breeding techniques will lead to or intensify extensive relocation of breeding products (breeding animals) from farms to commercial companies. This would result in a situation in animal breeding which is similar to that in plant breeding, where there is an inverted pyramid based on a few breeding companies, with a large number of propagating firms and numerous production firms. The pressure on the responsible EU organs and national governments to create a favourable environment for the commercial use of genetic and cloning technologies will probably increase. This includes for example demands to abolish so-called »competition-distorting regulations« such as quotas and ceilings on promotional measures or stocks. The consequences of possible further encouragement to this trend, which has been apparent in agriculture since the 70s, are not only quantitative but also qualitative. It would be important for policy-makers to ensure that a situation does not arise where the possible adverse effects of specialisation in animal breeding at the various levels of breeding, the labour market and farm structure are exacerbated or may even become irreversible. Overall, the application of cloning using nucleus transfer in agriculture requires careful consideration of the advantages and disadvantages. As far as the quantity and quality of human food is concerned, there is no direct need to clone animals for agricultural use. In addition, the currently foreseeable effects on the individual farm animal – and also on populations (breeding stock), breeds and possibly even species – are perhaps at least as serious as the impact on agricultural structures and the socio-economic conditions for people working in agriculture.

In ethical terms, an assessment of animal cloning must in principle be based on the same criteria which are (or should be) used in traditional animal breeding. In this respect there are various calls for the creation of a national ethics commission to deal with the moral and ethical issues of advances in biological and biomedical technology generally and with the consequences of advances in non-human biology and medicine. Its task would be to advise policy decision-makers and inform and educate the public. Possibly, intensive cooperation in the human area may be desirable with a national ethics commission which is also under discussion, and under certain circumstances a single ethics commission dealing with the entire human and non-human area of scientific and technological developments in biology and biomedicine may be meaningful. As broad groups of the public are also concerned about the use of cloning techniques with humans, the implementation of participative processes is important in opinion forming and policy advice (such as consensus conferences or citizens' fora). The need for forward looking and timely consideration and possibly legislation on cloning in humans cannot be lightly dismissed. Analysis of the legal aspects of animal cloning showed that this is currently legally permissible and subject to only limited restrictions under present legislation. In principle the question of animal protection is partly in contradiction to the actual goals of the Animal Protection Act in Germany at the present time, and in principle animal protection is secondary to the constitutionally-guaranteed freedom of research.

A different point of view and resulting implications could arise if the Constitution is amended to incorporate animal protection as a state goal. Constitutionally-based animal protection would presumably have relatively great practical effects. Such a provision would force the courts to strike a balance continuously between opposing constitutional rights and state goals. Developments in the field of cloning could in principle demonstrate the need to give animal protection constitutional status, or even the need for a fundamental debate in politics and society generally on the (current) treatment of animals in research and agriculture.

Captive Breeding & Species Reintroductions

Captive breeding and subsequent re-introduction of a threatened species is an important and in some cases very successful tool for species conservation. Critics point to the need to conserve/restore habitat, list examples of failures, decry the cost, and argue we should rescue species *before* they are on the brink of oblivion. Fair enough. But, captive breeding saved the bison. Wolves roam Yellowstone and the Upper Peninsula of Michigan, the Peregrine Falcon is off the endangered species list, golden-lion tamarins thrive in the Brazilian forests, whooping cranes perform their mating dances along river banks in the west, and many more species might similarly be rescued. Zoos, botanical gardens and aquaria have found new purpose and direction, providing a safety net when other protective measures have failed.

Terms

Ex situ conservation: captive breeding, gene and seed banks, zoos and aquaria and all other forms of maintaining species artificially and off-site. Contrasts with in situ methods such as parks and habitat management.

Introductions: releasing animals (captive or wild born) where they never existed. Usually because old habitat is gone or degraded, not available, but the new habitat is considered suitable.

Reintroductions: releasing captive born animals where they once existed. Only successful after you have corrected the cause(s) of the original population decline.

Translocations: moving wild-born animals from one place to another. This is done when the wild population is in imminent danger of extinction due to habitat alteration. One of Michigan's three populations of endangered redside dace (Clinostomus elongatus) was about to be wiped out by the installation of a new sewage treatment plant which would discharge lethal levels of ammonia into the section of the small creek where the fish live. In a last-ditch effort, a few concerned scientists gathered some fish and translocated them to Flemming Creek in the Botanical Gardens, where a small population established successfully.

Captive Breeding Programs

These programs arose out of the coincidence of two forces — unplanned parenthood by zoo animals raised the issue of what to do with surplus (zoos often had to destroy surplus animals); and concern for extinctions in the wild.

Although some species can be very hard to breed, captive breeding has a high success rate.

- 19% of all mammals, 10% of all bird species have been bred in captivity.
- 90% of all mammals, 74% of all birds added to U.S. zoo collections since 1985 were born in captivity.
- some species are extinct in the wild but thrive in zoos: Przewalski's horse, Arabian Oryx, Pere David's deer
- A number of wild populations of species were born in captivity and now live free: Bald Eagle, Golden Lion Tamarin, Andean condors, red wolves.
- A successful captive breeding program by US F&W with a bobwhite quail generated the creation of a wildlife refuge in southern Arizona to allow its successful reintroduction.
- public awareness and concern can be mobilized by such efforts.

Criticisms of Captive Breeding

Despite these statistics, captive breeding has its critics.

- It focuses on a few, charasmatic endangered species
- genetic diversity may have sunk too low to be regenerated
- costly, diverts resources from much more cost-effective ecosystem and habitat conservation measures
- gives false sense that battle against extinction is being won
- what if the habitat no longer exists, or cannot be expected with any confidence to exist in 200 years time. What point then ?
- NRE 220 students often react negatively to the degree of handling/intervention associated with captive breeding, and apparently would prefer to see species go extinct. Really?

Major Players in Ex Situ Conservation:

Zoos : Once places for people to stare at "curiosities", Zoos today are centres of captive breeding and opportunities for public education to heighten awareness about endangered species.

Game Farms : Viewed as repugnant by some, these propagate game species in semi-natural captive settings and reduce hunting pressure on wild populations. Excess animals produced may be source of genetic material for other breeding programs or may be candidates for re-introductions.

Aquaria : Fish and marine mammals. Mammals trained to perform for entertainment of public help to bring in money, prevents boredom in these intelligent animals, and today aquaria are important for public education. Some success with captive breeding of fishes, smaller marine mammals (esp. dolphins, manatee), less success with others (Orcas, sea lions and seals).

Captive breeding programs : set up specifically for the captive breeding of target species; animals not for public view. Smithsonian maintains a wildlife game farm inVirginia countryside.

Botanical Gardens : focus on ornamental and/or horticultural species.

Arboreta : traditionally focused on all sorts of trees, many ornamental and/or exotic. Problems is few trees of each species per arboretum.

Seed Banks : originally for the preservation of unique cultivars of horticultural and agricultural species. Primarily started for preserving unique, local cultivars of corn, rice, wheat, and potatoes during the "green revolution", when farmers started to abandon traditional, local cultivars in favour of high-yield hybrid strains.

Some new tricks boost genetic variability, increase the effective population size, and improve reproductive success.

1. Cross-fostering-Some animals can breed successfully in captivity, but are not able to raise their own young. In these cases, the young are raised by another (closely-related) species. Typical of birds. An example is Asiatic jungle fowl chicks incubated and reared (taught to forage, peck, scratch, and drink) by domestic hens; eggs of whooping cranes placed in nests of wild, closely-related sandhill cranes and raised by them.

2. Artificial Incubation-Also called the "head start program", used extensively with aquatic organisms and some birds which experience extremely high mortality during the critical hatchling period. Eggs are collected, incubated under artificial conditions, and reared to beyond the critical size or stage prior to release. (esp. useful for turtles, some fish).

3. Artificial Insemination-Sperm is collected from donor males, processed, and frozen for long-term storage and for worldwide distribution to inseminate females.

4. Embryo Transfer (mammals)-A single female can only have a limited number of offspring during her lifetime-e.g., a milk cow with high production can only have about ten calves during her lifetime, even though she has thousands of egg follicles in her ovaries.

The solution:

1. female is superovulated (hormonal induction of release of multiple eggs), and the eggs are collected via a surgical procedure called laparoscopy

2. Eggs are fertilized in a test tube with sperm from any of several donor males.

3. Embryos can be frozen for long-term storage, or immediately (or eventually) implanted into the uterus of a closely-related but non-endangered species for full-term development, birth, and subsequent foster rearing by the surrogate mother.

Fish Hatcheries: an "Evil" Twin and a "Good" Twin?

There is a long history of propagation of fishes in hatcheries for commercial and recreational fisheries, of course, and many view the success of these efforts from a conservation perspective as decidedly mixed. Hatcheries typically are used to supplement natural production of harvested species, and their goal usually is mitigation of other human activities that have led to declines in natural production (NRC 1996). Our appreciation of the negative aspects of hatchery propagation have been greatly clarified by examination of the health of the 7 species of Pacific Salmon occurring in western North America. The fact that many valuable stocks are in decline, despite rising output of hatchery-reared juveniles, shows that hatcheries alone cannot compensate for the problems of habitat loss, over-fishing, and dams and other barriers. Furthermore, because of lack of attention to genetic concerns, and the over-fishing of native populations in mixed fisheries of wild and reared fishes, hatcheries have done actual harm.

Consideration of ex situ conservation reminds us that genetics is of fundamental importance to the long-term conservation of species. Healthy species normally contain much genetic diversity, both within local breeding populations (demes) and between demes (NRC 1996). Genetic diversity represents the basis for adaptive evolution, and is evidence of adaptation to local conditions by individual populations. Loss of this diversity, through extirpation of local populations, fragmentation of previously inter-connected populations, and careless selection of breeding stock for hatcheries represent serious threats to long-term species survival (NRC 1996).

Genetic risks associated with hatchery propagation can be grouped into four classes: loss of within-population variability, loss of among-population variability, domestication, and extinction (Busack and Currens 1995). Loss of genetic variability within populations stems mainly from using too small a hatchery broodstock and from inappropriate mating protocols, potentially resulting in inbreeding depression, genetic drift, and artificial selection (National Research Council [NRC] 1996). These problems are almost certain to confront efforts to culture threatened species as well. Hatchery propagation contributes to loss of among-population variability mainly through the release of reared fish from nonindigenous broodstock into areas where local populations may persist, a practice now much less common (NRC 1996). Domestication, or genetic adaptation to hatchery conditions, can result from nonrandom selection of broodstock as well as from differences between hatchery and natural environments. Ironically, these effects of hatchery propagation together can contribute to the extinction of populations they are design to rescue.

Despite these considerable concerns about fish hatcheries, they potentially could play an important role in the conservation of threatened species. The hatchery infrastructure of the USA, and accumulated knowledge of scientists and hatchery operators, represents a formidable infrastructure for the propagation of threatened and endangered species. Re-direction of hatchery efforts towards the culture of imperiled species is a relatively recent change that holds the potential to aid in the recovery of many species, provided that it takes place within a holistic program that includes habitat protection. At present, 33 National (USA) Fish Hatcheries (out of a total of 74) are working with 28 federally listed fish species and 11 State-listed species in accordance with recovery plans and to prevent further declines and listings.

Dexter National Fish Hatchery, a warmwater facility in New Mexico, has been at the centre of intensive efforts to protect native fishes of the southwestern USA; to date, some 24 taxa of mainly riverine and spring-dwelling poeciliids, cyprinodontids and cyprinids have been held at this facility (Johnson and Jensen 1991). In addition, several rare taxa of salmonids in the genera Oncorhynchus and Salvelinus are held at coldwater facilties. The Dexter facility has pioneered a number of hatchery methods that may be less important in rearing sports fish, but are important when dealing with native species. These include once-through rather than recirculated water, to minimize diseases, strict measures to prevent mixing of populations, with the possibility of interbreeding, and extra attention to minimizing escape of fishes into nearby waters, which might establish non-native populations (Johnson and Jensen 1991).

An augmentation plan for the razorback sucker *Xyrauchen texanus* in the Upper Colorado River Basin (Modde et al. 1995) is a good example of how ex situ conservation efforts can aid in the recovery of an Endangered Species (USFWS 1991). At present only older (mostly 30-40-year-old individuals), non-recruiting adults persist in isolated segments of the fish's historic range. (Minckley et al. 1991). Artificial propagation, along with habitat rehabilitation and flow management, form the basis of a multi-agency plan to reconcile water development and razorback sucker recovery (Wydoski and Hamill 1991). Previous releases of fingerling razorback suckers (lower Colorado River: 1981-87; middle Green Riber (1987-89) were unsuccessful and indicate that stocking alone is likely to be insufficient. The present plan makes use of careful analyses of local riverine stocks (termed genetic conservation units, or GCUs), a cross-breeding strategy to maintain genetic diversity, along with flows to inundate bottomlands and better management of these nursery sites.

Development of rearing facilities for the razorback sucker exemplify the potential for innovative approaches to the propagation of endangered species. Initially fingerlings were reared in earthen ponds (Minckley et al. 1991). More recently, seasonally occurring backwaters along Lake Mohave's shoreline have been utilized and then made more permanent, and water heaters have been added to cold-water trout hatcheries to make them suitable (Mueller 1995).

Future Need for a Millenium Ark

Michale Soule likens our situation to Noah's. A catastrophic extinction event is imminent, driven by the rise in human population and technology, and the usurping of wildlands. Perhaps 20-25 % of the world's species will go extinct. How can we use captive breeding programs as the equivalent of Noah's ark ? Let's break this down into three questions: how long a voyage, how many staterooms, how many passengers ?

How long a voyage: how long will it be before habitat in the tropics begins to *increase* rather than decrease ? Best guess if plus or minus 500 years, barring catastrophes such as nuclear warfare. However, we can hope that cryogenic and similar technologies will be operational within 200 years — so that is the time line.

How many staterooms: how many species will require captive maintenance and propagation ? Best guess is about 2,000, excluding fish, and probably plus or minus 500. { This is roughly 10% }

This will place great demands on available space and resources, given that we must maintain large enough N_e to maintain essential genetic variation.

How many passengers: What population size is necessary to prevent the decay of genetic variation and ensure the future survivability of populations ? For animals that are very long-lived, an N_e of about 40 should do; for short-lived species we need more. Detailed genetic and demographic analyses are needed for each species to be maintained.

Bee Breeding and Animal Breeding

Most of these living forms have been cared for and propagated by man since long before the beginning of written history. Domestication implies several things, no one of which is sufficient to define it completely. It usually means tameness; but individual bears, lions, or even snakes can be tamed, and few of us would call them domesticated animals. Domestication further implies bringing the growth and reproduction of the plant or animal at least partly under man's control, but certain pigeons and geese usually choose their own mates and accept no others. Man uses domesticated animals or their products for his own advantage, and he usually keeps them in or near his dwelling. From the variable wild species man has, by selection and reproduction, developed individuals that suit his purpose. Breeding and selection have therefore led to the establishment of types or varieties useful to man. Selection for usefulness and survival under man's care has frequently resuited in varieties incapable of survival without man's care. Man did not create these genetic forms. By controlled breeding and selection he merely succeeded in putting together or fixing these varieties from the genetic variability that existed in their wild ancestors.

The honey bee has a definite place in our modern world. Its products of honey and wax are useful to man, although perhaps not essential to all men. However, the pollination activities of bees affect the lives of most of the people of the world. Although we may never succeed in taming the honey bee or in getting it to change its mating habits, it seems evident that the definition of domestication is sufficiently broad to include this most useful insect. The ways and means of improving the economic value of honey bees by controlling their growth and reproduction will be discussed in this article.

The breeder of animals must decide which animals shall produce the next generation. His choice is determined by his knowledge of the variability that exists in the potential breeding population and the usefulness that he desires in the succeeding generations. The variability in the breeding population is due in part to differences in inherited factors and also in part to differences in environment. A knowledge of the science of heredity should make it easier to choose intelligently among the many animals in the potential breeding population.

Because of the social nature of the honey bee, the colony is the unit upon which selection must be based for most economically important characteristics. The individuals of a colony consist of a queen and her worker and drone offspring. The queen is mated before she begins egg laying, and her mate is thus the father of the workers in the hive. She may have more than one mate, and therefore the genetic variability of the colony may be due to several individuals. Each of these individuals can be influenced by environmental factors that are not always easy to evaluate. The bee breeders' problems in properly evaluating the genetic and environmental factors affecting each individual colony are much more difficult than those of other animal or plant breeders.

Although honey bees have been propagated by man for 5,000 years or more, there is little evidence that much progress in bee breeding has been made. No superior breeds of bees have been established. The species has been subdivided into races, and all these races show considerable variability. Apparently they are the product of many generations of random matings within populations that have been partially isolated by geographic barriers. Man may have had some influence in fixing these races by selection that altered the frequency of certain inherited factors.

The aim of the bee breeder is to produce better or more profitable bees. This may mean more honey per colony or it may mean more efficient pollinators. The breeder must work with the reproductive individuals that are available. Since colonies of bees differ in many characteristics, the breeder has variability from which to select. This variability may be due to both genetic and environmental factors. The successful bee breeder must observe his colonies closely so that he can make proper allowances for environmental factors affecting the genetic variability in his breeding stocks.

His next problem is to mate the breeding individuals so as to obtain genetic improvement. From the variable colonies he attempts to unite the genes (the genetic factors) for good qualities from many stocks into one line or breed while eliminating inferior or less desirable qualities. In order to succeed he must have some knowledge of the science of heredity. Geneticists have accomplished this with many plants and animals, and therefore it is possible with bees.

Since the honey bee is an animal, it would appear that animal breeding methods should be used. The honey bee, however, is quite different from the economically important animals. Its method of reproduction is in a sense more like some plants. We shall attempt to describe these differences and likenesses and show how they affect methods of bee breeding.

The science of heredity is relatively new, having only become recognized as a science in the 20th century. Yet a great amount of real progress through selective breeding was accomplished in most of our economically important plants and animals long before Mendel, the father of genetics, was born. Most every breed of horses, cattle, hogs and chickens used in the United States today was established before the 20th century. Yet even today there are no fixed breeds of bees – we still have only races.

Let us review a few of the methods employed in the establishment of improved breeds of livestock. The fundamental biological principles of inheritance are the same whether we are breeding plants or animals. Therefore, much about bee genetics can be learned by studying the genetics of other animals and plants.

All our domestic animals except the guinea and the turkey originated and were domesticated in prehistoric times in the Old World. Most domestication evidently occurred in central or western Asia, although some is believed to have taken place in Europe, Egypt, India and China. It is still disputed whether most animals descended from a single wild species or from two or more species.

Early naturalists supposed that species were fixed and had no genetic variability. However, modern geneticists have shown that wild populations of one species of animal were not genetically uniform. They have also shown that selected breeding within these wild populations, with their existing genetic variability, can produce distinctly contrasting races or breeds within a few generations.

It seems probable that all the races of bees, as we know them today, developed from a single wild species of honey bee. In time they became dispersed over a large part of the Old World and were somewhat isolated into many small groups. There was, of course, some interbreeding between groups, but geographical barriers prevented a great many matings between groups. As a result a certain amount of inbreeding took place within each geographic group. By mutations, natural selection, and perhaps some human intervention in selection, the bees within each geographic area became different from those in other areas. The bees from the various areas are called races. Probably the most geographically isolated race of bees is the Caucasian. A Russian author who studied this race in its native home divided it into six separate varieties. These varieties ranged in colour from the banded appearance of the Italian to completely black. Other genetic characteristics within this and other races of bees are equally variable. The bees of one race are fertile when crossed with those of another race.

Wild cattle, horses, and hogs were also variable animals. The recognized breeds of these animals were developed by selected breeding of outstanding individuals from variable populations. The history of the formation of the beef breeds of cattle that originated in the British Isles is typical of successful animal-breeding methods.

History tells us that migrating people and invading armies usually carried with them much livestock from their native lands. Migrations were slow, and the stocks mingled with and interbred with stocks present in the countries through which they passed. It is not known just how many and what kinds of cattle were introduced into the British Isles throughout the ages. However, it is known that the Norsemen and the Dutch carried cattle into the British Isles, and it is suspected that the Romans and the Normans also brought in some cattle. These cattle interbred with the cattle of the islands, and a great amount of genetic variability existed in these stocks before the formation of the breeds that we know today as Hereford, Shorthorn, and Aberdeen-Angus.

About the beginning of the 18th century the common lands of Great Britain were enclosed so that the intermingling of stocks on a wide scale was stopped. The cattle population was thus broken up into small groups and inbreeding within groups was the result. About this time the cities began to grow and a demand for meat developed. Some types of cattle were more useful for this purpose than others, and breeders attempted to produce these meat-type cattle. A few breeders gathered together some of the best of the desired type into one or a few herds. There followed years of rather intense inbreeding between these selected animals and their descendants. Very little outside breeding stock was permitted to enter these herds. After several generations the cattle within these herds were uniform and were distinct from the other animals in the community. Every recognized breed of cattle has arisen from the concentration of the genes (blood) of one or a few animals of greater than ordinary merit.

Breeds have developed in a similar manner among other types of livestock. The Poland China hog was evolved or originated in Butler and Warren counties in southwestern Ohio. The exact combination of breeds or types that resulted in this hog is not clear. Prior to 1830 there existed in this area a number of types, among which were the Bedfordshire, Byfield, Russian, and Big China. Some Berkshire blood was introduced in 1835 and some Irish Glazier blood in 1839. There was intermixing of these breeds without a definite plan for 20 years or more. In 1870 the Warren County hog was black and white spotted. It was regarded as a breed in 1872 and was given the name "Poland China." The diverse races of swine from which this breed was found could hardly have come together anywhere on earth without the aid of man. From this heterogeneous mixture the breed arose by inbreeding of selected sires.

Why have not bee breeders done the same thing with bees? Many diverse races and strains from all areas of the world have been brought into the United States. There is thus great genetic variability in our bees. When we look around we find outstanding individuals (colonies). We have seen that successful breeders of other animals succeeded in establishing superior breeds of animals by inbreeding and selecting from among outstanding individuals and their descendants.

Bee breeders did try methods used by breeders of other animals but they did not succeed. It is true that they could not control matings so successfully as other animal breeders, but this was not the major cause of their failure. Studies indicate that a major cause for this failure was a lack of knowledge of sex determination in honey bees. Its importance to bee breeding will be explained in a later section.

Patentability of Animals

Industrially useful inventions are generally patentable if they are novel and involve an inventive step (i.e. they are surprisingly different, rather than simply a routine development, modification or combination of previously known features). Biotechnology is providing the capability to produce novel genetically modified organisms (plants, microorganisms and animals) which are commercially valuable and industrially useful. Patent protection for these innovations is being sought in many countries.

When a patent is granted, the monopoly right given by the patent is defined in patent claims which must be supported by a detailed description with one or more examples of how the invention can be carried out and used. The description may describe variations on the basic inventive concept covering a number of ways in which the invention can be put into practice.

The grant of a patent confers only a civil right on the patent owner to prevent others doing what is claimed in the patent. It does not give the owner any specific rights to exploit the invention himself. He might be prevented from exploiting the invention by virtue of a patent right held by a third party, or he might be prevented by reason of law or regulation. For example, an inventor might successfully obtain a patent for a particularly unusual construction of a rocket-propelled motorcycle, but in order to use it he would have to satisfy other laws and regulations relating to the control of road vehicles. If such vehicles were completely forbidden for use on roads it would nonetheless not prevent him from experimenting with them, obtaining patents for them and testing them under controlled conditions. If the invention was commercially valuable but no patent protection was available that might not, of itself, stop him from developing the technology but it might deter others from investing money in any business set up to exploit the technology of the invention. There would be no incentive to put the effort into development and commercialisation if competitors could copy the finally developed idea. These same considerations apply, particularly with respect to regulation and control, in the field of patenting biotechnology.

By specific provisions of the UK and European patent laws it is not possible to have a valid patent claim which claims an essentially biological process as such (e.g. a method comprising mating a bull from one cattle breed with a cow from another cattle breed to produce a cross-bred calf). Nor is it possible to have a valid patent claim with scope so narrow as to cover only a group of animals comprising an "animal variety" as such. An "animal variety" is not legally defined in European patent law, but may be taken to be a group of animals of the same species which have been selected to constitute a breed having at least one significant and identifiable characteristic. The meaning of the term "breed" is well defined and understood within the farming industry. Patents should not be granted for inventions which are judged morally offensive or against the "ordre public". There is no absolute criterion of moral offensiveness: the decision rests with the Patent Offices and Courts in each country and, ultimately, with public opinion.

Recent advances in animal husbandry and biotechnology mean that new organisms can be bred which differ from previous organisms of the same type by virtue of a modification caused by human technical intervention. Hence the production of genetically modified organisms, whilst possibly being for the most part an essentially biological process, may now involve a critical process step which is essentially non-biological (e.g. the insertion of a segment of a foreign gene into the animal's DNA). Such processes (being a combination of essentially biological steps in combination with other steps) taken as a whole fall outside the specific exclusion clauses and are therefore patentable provided they meet all other requirements for patentability. Furthermore, the new techniques and the new products for which patent protection is sought are generally applicable or obtainable over a whole range of species or even genera. The inventor of a useful and advantageous modification should in principle be able to obtain patent claims of broad scope which would cover any animal from a broadly defined group (larger than a breed or variety) within which all group members embody the invention. Claims to such a group fall outside the specific exclusion of claims to animal varieties.

In Europe, the Biotechnology Directive 98/44/EEC (approved by the European Parliament in May 1998) allows the patenting of animals (subject to certain conditions). The test case in Europe is "The Harvard Mouse" or "oncomouse" patent application filed on 24 June 1985. The inventors devised a technique for inserting a gene which increases sensitivity to cancer (an oncogene) into the DNA of animal cells and this enabled the inventors to clone laboratory test animals (generally mice) which are useful for testing and identifying any potential cancer-causing chemicals. After long and detailed examination by the EPO Examining Division and Board of Appeal, the case granted on 13 May 1992. The granted patent included claims to the technique and also to non-human animals produced by the technique. In allowing the case, the EPO considered both the technical questions of patentability and the numerous ethical questions raised by many third party objectors.

However, the matter was not closed because there was opportunity for objectors to oppose the grant of the patent by the EPO. Oppositions were indeed lodged by a number of objectors including a coalition of animal welfare groups. All of the oppositions were lodged on ethical/moral principles (not on technical grounds). A final hearing was held in November 2001, and a decision from the Opposition Division of the EPO issued on 16 January 2003. The Opposition Division decided to maintain the granted patent, with the claims restricted in scope to cover the invention carried out in rodents only (rather than any non-human animal as per the originally granted claims). The patent claims upheld by the Opposition Division covered methods of making transgenic rodents and transgenic rodents per se, as well as directly related plasmids, chromosomes, cells, methods of culturing transgenic rodent cells, and methods of testing for carcinogenic materials.

Several of the opposing parties filed Appeals in March 2003 against the Opposition Division's decision (Appeal Reference EPO T0315/03). At Oral Proceedings on 5-6 July 2004, the Appeal Board decided to uphold the patent in a further restricted form in which the term "rodent" was limited to "mouse". Reasons for the Appeal Board decision were published by the EPO in March 2005. The Appeal Board considered two key issues in detail: (i) whether in the present case the process for modifying the genetic identity of the animals would be likely to cause the animals suffering without any substantial medical benefit to man or animal; and (ii) whether the publication or exploitation of claimed invention would be contrary to "ordre public" or morality. The Appeal Board held that they could see no medical benefit in relation to rodents as a group (including species such as squirrels) but that the invention when limited to mice was shown to have substantial medical benefit. The contrasting positions put forward on the issue of morality reflected divergent European perspectives, but the Appeal Board held that the usefulness of invention to mankind was sufficient to outweigh the evidence before it against "ordre public" or morality. By the time the Appeal Board's final decision was published, the upheld "Harvard Mouse" patent was only a few months away from its 20 year patent expiry term. However, the oncomouse test case has confirmed that in principle it is possible at present to obtain a European patent for a non-human animal produced by genetic modification.

Both the US and Japan have already recognised the patentability of animals.

In contrast, the corresponding "Harvard Mouse" case in Canada has been held to be unpatentable because Canadian patent law excludes higher life forms.

Man has always owned and exploited animals for his own purposes both as pets and as livestock. Animals are goods of trade and fall within the definition of "goods" in the UK Sale of Goods Act 1975. The genetic modification of domestic animals has been carried out to-date by traditional breeding methods and selection. Legislation exists in the UK and Europe to control genetic experimentation (such as Directive 86/609/EEC regarding the protection of animals used for experimental and other scientific purposes), and to control the introduction of genetically modified organisms into the environment (such as Directive 90/219/EEC and Directive 90/220/EEC regarding the contained use and deliberate release into the environment of genetically modified organisms). The general public must be reassured and have confidence that the necessary rules and safeguards are in place and effectively enforced. In the UK, genetic experimentation is strictly controlled by the Health and Safety Executive. So long as innovative research and development in the field of biotechnology continues to be accepted as part of technological progress, results of such efforts should be protectable by intellectual property rights, provided that the criteria for patentability are met. European patent law now provides certain additional criteria for assessing whether or not a process for genetically modifying animals, or animals produced by such processes, are patentable, for example by imposing a test which weighs up the likelihood of suffering by the animal against any substantial medical benefit to man or animal.

Introductory Population Genetics

Population genetics: gene and genotypic frequency. Hardy-Weinberg Law, forces affecting gene frequency and composition of population, effective population size. Quantitative genetics: quantitative characters and their inheritance; biometery and its importance in animal breeding, biological population and its parameters; statistical methods in animal production: measures of central tendency and dispersion for quantitative traits. Basic theorems of additivity of variance, regression and correlation concepts. Experimental designs. Nature and causes of variation in quantitative traits and partitioning of hereditary variance. Concepts of heritability and repeatability and methods of their estimation. Genetic and phenotypic correlations. Qualitative inheritance: genetic basis and causes of variation. Lethal and semi-lethal genes, autosomal and sex-linked lethals, effects on sex ratios, deletion and elimination of lethals from livestock populations. Other genetic abnormalities. Introduction to computers: software and hardware.

Exercises on gene and genotypic frequencies in random mating populations. Numerical exercises on the estimation of various statistical parameters and hypotheses testing using computers. Exercises on the estimation of heritabilities, repeatabilities and genetic correlation from given data. Case studies for lethal, semi-lethal and other genetic abnormalitiesin farm animals.

Principles of Animal Breeding

Origin and domestication of farm animals, consequences of domestication, breeds of livestock and their development, role of breed registry societies/associations in developed countries and its application in Pakistan. Selection: Natural and artificial selection, methods of selection; tandem method, independent culling level and selection index. Kinds of selection; mass selection, selection based on multiple records or lifetime average, pedigree selection, progeny testing, family selection. Selection for single and multiple traits, correlated response, genetic effects of selection, methods of assessing genetic progress. Various systems of breeding; random mating; inbreeding and its effects on small and large populations.

Measurement of coefficient of relationship and inbreeding; line-breeding for increased prepotency, outbreeding; outcrossing, crossbreeding, grading up, phenotypic assortative matings. Development of inbred lines, selection for best combining abilities, reciprocal recurrentselection. Breeding for threshold characters.

Estimation of genetic gain; evaluation of livestock on the basis of their own performance, pedigree and progeny. Calculation of breeding values from single and repeated records. Measurement of coefficient of relationship and inbreeding by methods of paths and variance-covariance chart. Measurement of heterosis. Estimation of genetic changes in performance traits due to various mating systems.

Selection For Economic Traits in Farm Animals

Traits of economic importance in dairy animals, genetic correlations among the economic traits, their heritability and repeatability values, genetic methods to improve productivity, selection of dairy heifers and bulls, use of standardized records, relative economic values, breeding values and selection indices. Traits of economic importance in beef animals, genetic parameters of these traits, feasibility of producing beef animals in Pakistan, crossbreeding for milk and meat production. Traits of economic importance in draught animals, correlation among these traits, methods of their genetic improvement, role of crossbreeding in draught animals. Traits of economic importance in sheep and goats, genetic correlation among these traits, their heritability and repeatability values, methods to improve productivity, female selection, selection of rams and bucks. Traits of economic importance in poultry and their improvement, formation of breeding stock for layers and broilers, development of dual purpose birds and rural poultry. Introduction to various statistical packages for genetic parameters and breeding value estimation.

Exercises on the maintenance and standardization of productive and reproductive records. Estimation of Expected Real Producing Ability (ERPA) and Breeding Value using standardized records.

Exercises on the estimation of genetic parameters. Calculation of relative economic values. Construction of selection indices for large and small animals. Calculating different genetic parameters through computers.

Genetic Aspects of Farm Animal Reproduction

Biological basis of sex. Normal and abnormal mechanisms in sex differentiation. Chromosomal aberrations and inherited abnormalities of the male and female reproductive systems. Genetic factors in fertility and sterility. Reproductive performance: inheritance of reproductive traits and their relationship with other economic traits. Reproductive efficiency, its scope and importance in farm animals, measures of reproductive efficiency in cattle/buffalo, sheep and goats and their computation. Genetic and environmental basis of the differences in reproductive efficiency. Augmenting reproductive rates, application of biotechnology. MOET schemes for increased genetic gains.

Microscopic examination of normal and abnormal chromosomal spreads for identification of different reproductive disorders. Study of the genitalia of farm animals for comparison of normal and problem cases. Semen evaluation with normal and abnormal karyotypes. Exercises on the estimation of reproductive efficiency.

Farm Practices

Identification of various types and breeds from phenotypic expression of the animals. Score carding, Selection of breeding animals for type and breed characteristics. Record keeping: requirements and formats for different species. Development and maintenance of history sheets and other registers. Data coding, entry and computer analysis. Comparison of different housing structures for comfort and productivity of various genotypes. Preparation of purebred and crossbred animals for show purposes. Measurement of physiological norms as adaptability indicators. Daily routine practices at the farm, hand and machine milking. Sire summary appraisals. Progress of the cross-and pure breeding projects, scope and limitations. Examination of genital organs of animals for normal and abnormal inherited traits. Usage and preservation of gametes for conservation of animal genetic resources.

Genetics and Animal Breeding

Physical basis of inheritance; gametogenesis, spermatogenesis, oogenesis. Probability: concept and laws of probability. Mendelian basis of inheritance, monohybrid, and polyhybrid segregation ratios. Multiple alleles, pleiotropy, modified segregation ratios, epistasis. Sex determination and sex linkage, sex limited and sex influenced traits. Linkage and crossing over. Chemical basis of inheritance, the gene, gene action, DNA and its replication, expression and interaction of genes, gene mutation and chromosomal aberrations, lethal and sub-lethal genes. Genetic basis of variation :phenotypic, genetic and environmental variance; heritability, repeatability and genetic correlations. Concept of gene and genotypic frequency, forces that change the gene frequency and genetic structure of a population: selection, mutation, migration and random drift. General principles of selection: Selection on the basis of individual performance, pedigree, family, progeny and all available information. Breeding systems based on relationships and phenotypes, pure-breeding, crossbreeding and grading up.

Laboratory demonstrations on cell division and gametogenesis, exercises on segregation ratios and probability. Exercises on calculation of coefficients of relationship and inbreeding. Record keeping and evaluation of individual, pedigree and progeny performance, selection exercises and assessment of genetic gain.

Animal Breeding Practices

Role of animal breeding in livestock production. Opportunities for breeding and improvement of farm animals in Pakistan. Conservation of animal genetic resources in Pakistan: scope, techniques and problems. National breeding policy for improvement and conservation of livestock. Constraints in improving the productivity of livestock under traditional breeding systems. Review of the breeding practices used by the developed countries for increasing the performance of farm animals. Future breeding plans for improvement of farm animals for increasing productivity in different agro-ecological zones of Pakistan. Emerging breeding technologies for increased animal productivity.

Computation of various productive and reproductive traits (calving interval, dry period, service period, productive life, herd life, fertility rate, hatchability etc.) in different farmanimals from the available records. Exercises on the feasibility of purebreeding and crossbreeding in farm animals for milk, meat, wool/hair and egg production under local conditions. Project preparation on specific topics. Use of internet databases on animal genetic resources.

Applied Animal Breeding and Internship

Requirements of identification and recording of farm animals for genetic improvement programs. Needs for data processing and evaluation at national level. Genetic progress simulation. Review of the current projects on livestock improvement in the country. Future animal breeding policy. Report writing. Visits to the livestock farms, livestock production research institutes, semen production units, poultry breeding farms and other related organizations for practical training under the guidance of a teacher as well as the staff of the institute/farm concerned. Submission of a concise report on the practical experience gained during the visit of these organizations. Practical training will especially cover the following:

a) Identification of animals: identification methods, coding of identification numbers for computer entry and analysis. Attributes of various breeds pertaining to economic importance and score carding.

b) Record keeping: Maintenance of history sheets and other registers for recording information on various species. Daily record keeping schedules.

c) Conversion of information from different registers and history sheets for electronic data storage, analysis and presentation. Use of computer packages for data handling. Selection of animals on the basis of individual, pedigree, progeny and other information.

d) Breed improvement work being done at various livestock stations on buffalo, indigenous and crossbred cattle, sheep and goat: Plan of work, performance and evaluation of progress achieved.

e) Rectal palpation for checking the condition of reproductive organs, heat detection and pregnancy diagnosis. Insemination of buffalo, cattle, sheep, goats and poultry. Calving and lambing forecast. Semen collection from bulls, its evaluation, dilution, preservation and storage. Management of bulls and rams used for semen collection.

General Genetics

The scope and significance of modern genetics, heredity and phenotype. Cytological basis of inheritance; the cell cycle. Sex chromosomes and sex determination. Mendelian genetics, modifications of segregation ratios, multiple allelomorphism. Linkage, crossing over and chromosome mapping. Genetic variation: variation in chromosome number and arrangement; mutation and mutagenesis; extra-chromosomal inheritance. The Chemical basis of heredity, the genetic material, nucleic acids' structure and analysis, replication and synthesis of DNA, organization of DNA in chromosomes, gene structure, function and regulation. Immunogenetics. Genetic engineering: Techniques for gene cloning; application of rDNA technology. Qualitative vs quantitative inheritance, genetic and artificial selection, gene frequency and forces affecting it; Control of quantitative characteristics; Darwinism and evolution; conservation of genetic resources.

Numerical exercises on problems discussed in theory. Laboratory exercises and analysis of breeding experiments with Drosophila/tribolium illustrating the laws of heredity. Demonstration of various cytogenetic techniques.

Population Genetics

Review of the basic concepts of genetics and statistics. Genetic structure of populations:

Hardy-Wienberg law and equilibrium; gene and genotypic frequencies. Gene counting methods, Maximum Likelihood estimation of gene frequency. Fisher's fundamental theorem of natural selection. Relationship and inbreeding coefficient for an autosomal locus; for a sex-linked locus. Consanguineous matings and rare traits inbreeding in a randomly mating population of finitesize, effective population size. Evaluation and uses of Hardy-Wienberg equilibrium, assumptions underlying Hardy-Wienberg equilibrium. Changes in gene frequencies: Mutation, migration, selection, random drift. Genetic basis of variation: Mode of gene expression, measurement of variation. Heredity and environment: Role of heredity and environment ininfluencing quantitative traits and its measurement. Genetic correlation and genotype environment interactions. Genetic integration of Mendelian populations, Bay's theorem.

Exercises on changes in gene and genotypic frequencies under different conditions.

Measurement of relationship and inbreeding. Estimation of variance components, Binomial and multinomial distributions. Simulation studies on Bays Theorem and Maximum Liklihood estimation procedures.

Animal Breeding Plans

Genetic principles in animal breeding. Heredity and environment, genetic parameters: heritability, repeatability and methods of their estimation; phenotypic, genetic and environmental correlations. Breeding plans based on selection, aids to selection, selection between and within populations. Mixed model solutions and Best Linear Unbiased Prediction (BLUP). Selection limits, selection intensity, accuracy of selection, selection index and correlated response. Estimation of genetic and economic gains. Optimizing genetic progress. Breeding plans based on relationship: Likeness between relatives and degree of relationship. Inbreeding, linebreeding. Outbreeding systems, rotational crossing; Crossing to produce a synthetic, grading up. Family structure of populations. Breeding plans based on somatic likeness. Nucleus system: Open and closed nucleus system, MOET. Simulation of selection process.

Problems on the determination of breeding value of an individual using lifetime average, pedigree information, progeny performance and other sources. Estimation of breeding values using Best Linear Unbiased Prediction (BLUP) procedures. Exercises on selection differential, genetic gain and selection indices.

2

Breeding of Farm Animals

Objectives in the breeding of farm animals, role of animal breeder in improving livestock economy. Existing livestock genetic resources with particular emphasis on production potentials; comparison with foreign breeds and basis for the superior/inferior performance. Breeding policies for cattle, buffaloes, sheep and goats and other livestock species in Pakistan. Strategies for improved use and conservation of livestock. Inbreeding and crossbreeding for livestock improvement: Development and uses of inbred lines in poultry. Heterosis: genetic aspects, exploitation, crossbreeding systems. Detection and elimination of genetic defects and lethals from the stock. Maintenance of performance records, their standardization and optimal use. Large scale genetic evaluation. Interpreting genetic evaluation information.

Practicals

Exercises on the evaluation of productive traits of various breeds of livestock. Exercises on standardization of records, genetic evaluation and selection, Computer usage for animal evaluation. Review paper required. Visits to different breeding farms.

Advanced Animal Breeding

Current animal breeding, estimation of genetic parameters, selection for economic merit, testing and selection within stocks. Role of A.I. and other related techniques in enhancing the genetic progress. Multi Trait Across Country Evaluation (MACE) and its application. Genetic technologies: role of biotechnology in animal production, recombinant DNA techniques, recombinant DNA for the improvement of domestic animals; Micro-manipulation: cloning, homozygous diploids, androgenesis and gynogenesis, sex selection, chimaeras, transfer of genes. The direct detection of genotype, DNA polymorphism. Expected genetic and economic gains. Detecting major genes, basic concepts of marker based analysis, influence of markers on production and reproduction in farm animals, mapping and characterizing Quantitative Trait Loci (QTL), genetic maps. Review of the recent literature in animal breeding.

Statistical Methods in Animal Production

Basic biometrical concepts and definitions. Data description for single and multiple variables. Probability and its distributions. Estimation and hypothesis testing. Concepts and application of regression and correlation, least squares analysis. Introduction to analysis of variance and assumption underlying it. Principles of planning of animal experiments.. Conceptsbehind using various experimental designs such as completely randomized design, randomized complete block designs and Latin square design with and without factorial arrangements. Changeover, crossover and other designs. Relative efficiencies of these designs. Multiple comparisons. Fixed, random and mixed models. Estimation of variance components under various statistical models. Categorical animal data analysis, animal production data management and report preparation.

Introduction to various computer packages for animal data handling and processing.

Numerical exercises on various experimental methods listed in the theory manually and by using computers.

Efficiency of Reproduction in Farm Animals

Genetic basis of variation in reproductive functions. Relationship of reproduction to genetic improvement in domestic animals; Pre-and post natal selection of genetic and reproductive material; the problem of reduced fertility and reproductive efficiency and its relation to lifetime productive efficiency; Measures of reproductive efficiency in farm animals; Genetic and environmental basis of the differences in reproductive efficiency; Fertility level of males for use in artificial insemination; Reduction in estrous period, quiescent heat and anestrus; Embryonic losses; Issues related to prolificacy; Manipulation of reproductive cycle, Feeding and management practices in improving fertility traits; Augmenting reproductive rate, Application of biotechnology for improving reproductive efficiency. Exploitation of genetic variation in reproductive performance for improvement.

Practice on the slaughter house material and experimental live animals. Rectal palpation of genital organs at various stages of reproductive cycle. Practice in various methods of heat detection; correction of reproductive problems. Hormonal application and other techniques. Practice in pregnancy diagnosis in various livestock species. Measurement of reproductive fitness in males. Determining reproductive efficiency in males and females.

Biometrical Genetics

Statistical concepts in animal genetics. Review of matrix algebra and linear models, elementary probability and its application to Mendelian inheritance. Metric characters, population mean, average effect, breeding value, dominance deviations, interaction deviations. Sources of genetic variation in single and multilocus traits, sources of environmental variation. Estimation of variance components: parent-offspring regression, sib analysis, cross-classified designs, genotype x environment interaction, maternal effects and other genetic models. Estimation of breeding values.

Theory of Inbreeding

The concept of inbreeding and its measurement. Wright's method of path coefficients and variance co-variance chart, Malecot's concept of relationship and inbreeding. Distribution, mean and variance of gene frequency in small populations. Genotypic frequencies with inbreeding. Regular systems of inbreeding. Inbreeding with two loci. Effects of inbreeding on the variance, effective population size. Dangers of inbreeding, inbreeding depression and heterosis. Practical uses of relationship coefficients. Selection and inbreeding, simulation as applied to inbreeding.

Experimental Techniques in Population Genetics

The lifecycle of Drosophila melanogaster, Genetics of embryonic development in Drosophila and Tribolium, different stages, methods of breeding and recording. Methods of genetic analysis with respect to gene action and properties of genetic material. Analytical procedures by means of markers. Verification of theories of population genetics on the colonies of the above species with special emphasis on quantitative traits like bristle number, larval weight, etc. Review of research work done.

Handling and breeding of Drosophila and/or Tribolium species. Experiments of gene frequency, genotypic ratios, Hardy-Weinberg law, and departures from it in small populations. Demonstration of selection and inbreeding experiments for verification of quantitative theory.

Poultry Genetics and Breeding

Basic genetic concepts. Poultry genetic resources, reproductive biology in relation to genetics and breeding, sex determination, genetics of the plumage, skin, growth, meat production, egg production, egg composition, fertility, diseases and behaviour. Genetic controls in selection, genotype by environment interaction, heterosis and inbreeding. Development of commercial synthetic layers and broilers, Molecular and immunogenetic aspects of poultry genetics, genetic engineering, mutations and major variants in the fowl. Breeding for fancy poultry, Breeding for rural poultry production systems. Conservation of poultry genetic resources.

New Animal Breeding Techniques and their Application

The new biotechnologies of gene transfer, in-vitro production, cloning and sexing of embryos have been developed and are being refined with efficiencies suitable for use in animal agriculture. Efficient in-vitro systems for maturing oocytes and capacitating spermatozoa, for fertilizing and developing the embryos have resulted in commercial in-vitro production of embryos. Cloning of embryos by nuclear transfer has been accomplished for sheep, cattle, pigs and rabbits with nuclear material supplied by embryos as late as the 120-cell stage in sheep. Embryos have been recloned but much research is needed to increase the efficiency of this procedure. Research is needed to develop the use of cultured cells in embryo cloning so that the number of clones may be increased to thousands or millions. Embryos of most species can be sexed in a non-damaging way with male specific antibodies and a more efficient method, amplified DNA hybridization, is beginning to be tested commercially.

Transgenic embryos or offspring have been produced for mice, rats, rabbits, chickens, fish, sheep, pigs and cattle. Genes can be targeted for expression in specific tissues but more efficient methods and a better understanding of the genes to be transferred as well as control by man of the time and tissue of specific gene expression are needed. Before many transgenic animals of value can be made, we must know which genes to introduce. Presently there is a poor understanding of the genes influencing animal growth, efficiency of growth, environmental adaptation, meat, milk or egg composition or animal disease resistance. Their identification will come from badly needed efforts to map the genome of domestic animals. These and other new technologies promise to change livestock breeding drastically in the next decade.

Breeding (Animal)

The application of genetic principles to improving heredity for economically important traits in domestic animals. Examples are improvement of milk production in dairy cattle, meatiness in pigs, feed requirements or growth rate in beef cattle, and egg production in chickens. Selection permits the best parents to leave more offspring in the next generation than do poor parents.

Selection is the primary tool for generating directed genetic changes in animals. It may be concentrated on one characteristic, may be directed independently on several traits, or may be conducted on an index or total score which includes information on several traits. In general, the third method is preferable when several important heritable traits need attention. In practice, selection is likely to be a mixture of the second and third methods.

Heritability, the fraction of the total variation in a trait that is due to additive genetic differences, is a key parameter in making decisions in selection. Most traits are strongly to moderately influenced by environmental or managemental differences. Therefore, managing animals to equalize environmental influences on them, or statistically adjusting for environmental differences among animals, is necessary to accurately choose those with the best inheritance for various traits.

The improvement achieved by selection is directly related to the accuracy with which the breeding values of the subjects can be recognized. Accuracy, in turn, depends upon the heritabilities of the traits and upon whether they can be measured directly upon the subjects for selection (mass selection), upon their parents (pedigree selection), upon their brothers and sisters (family selection), or upon their progeny (progeny testing). For traits of medium heritability, the following sources of information are about equally accurate for predicting breeding values of subjects: (1) one record measured on the subject; (2) one record on each ancestor for three previous generations; (3) one record each on five brothers or sisters where there is no environmental correlation between family members; and (4) one record each on five progeny having no environmental correlations, each from a different mate.

Propagation of improved animal stocks is achieved primarily with purebred strains descended from imported or locally developed groups or breeds of animals which have been selected and interbred for a long enough period to be reasonably uniform for certain trademark characteristics, such as coat colour. Because the number of breeding animals is finite and because breeders tend to prefer certain bloodlines and sires, some inbreeding occurs within the pure breeds, but this has not limited productivity in most of these breeds. Crossbreeding makes use of the genetic phenomenon of heterosis. Heterosis is improved performance of crossbred progeny, exceeding that of the average performance of their parents. Most commercial pigs, sheep, and beef cattle are produced by crossbreeding.

Advances in a variety of technologies have application for improvement of domestic animals, including quantitative genetics, reproductive physiology, and molecular genetics. Quantitative geneticists use statistical and genetic information to improve domestic animals. Typically a statistical procedure is used to rank animals based on their estimated breeding values for traits of economic importance. The statistical procedures used allow ranking animals across herds or flocks, provided the animals in different herds or flocks have relatives in common. The primary contribution of reproductive physiology to genetic improvement is to reduce the generation interval. If genetic improvement is increasing at the same rate per generation, more generations can be produced for a fixed time, and thus more gain per unit of time. The most important development was artificial insemination, which allows extensive use of superior males. Another development was embryo transplantation, which allows more extensive use of females. Cloning is a relatively new technique, by which whole and healthy animals have been produced that have the same DNA as the animal from which the cells were taken.

Due to advances in molecular genetics, knowledge is increasing regarding the location of genes on chromosomes and the distance between the genes. In domestic animals, polymorphisms (changes in the order of the four bases) that are discovered in the DNA may be associated with economic traits. When the polymorphisms are associated with or code for economic traits, they are called quantitative trait loci (QTL). When a few or several quantitative trait loci are known that control a portion of the variability in a trait, increasing the frequencies of favourable alleles can enhance the accuracy of selection and augment production. Another use of molecular genetics is to detect the genes that code for genetically predetermined diseases.

An example is the bovine leukocyte deficiency gene, which does not allow white blood cells to migrate out of the blood supply into the tissues to fight infection. The calves perish at a young age. Screening all sires that enter artificial breeding organizations and not using sires that transmit the defect has effectively controlled this condition.

Genetic Engineering

The last four years of the twentieth century witnessed the most rapid adoption of a new technology in history. Since 1996, millions of acres of farmland have been planted with genetically engineered (GE) crops—mainly corn, soybeans, and cotton. This means that genetically engineered organisms (GEOs) are being released to the environment on a massive scale, an event unprecedented in the 3.8 billion year history of life on this planet. This technological upheaval happened virtually without public debate, while our government played the role of enthusiastic promoter, rather than cautious regulator, of this radically new and environmentally hazardous technology.

Genetic engineering is a new technology that can combine genes from totally unrelated species in ways not possible using conventional breeding methods. Genes from an animal, say, a fish, can be put into a plant, a strawberry for instance. An attempt to "improve" strawberries by inserting a gene from an Arctic fish has in fact been discussed. The fish gene is supposed to make the strawberries more resistant to frost by causing the strawberry plant to produce a form of antifreeze which the fish normally produce to endure cold ocean conditions.

Over 60 percent of all processed foods purchased by U.S. consumers are manufactured with GE ingredients. Some corn and potatoes have even been genetically engineered to contain a gene from Bt bacteria which causes every cell of the plants to produce an insecticidal toxin. Yet there is no labelling of these or any GE foods as being genetically engineered, because the U.S. Food and Drug Administration (FDA) considers the GEOs from which these foods are made to be "substantially equivalent" to the non-genetically engineered plant from which the GEOs are derived.

The doctrine of substantial equivalence is pure pretext and rationalization with no basis in science. Yet, in a remarkable display of arrogance, the supporters of genetic engineering accuse their critics of not basing their objections on "sound science."

The FDA also uses this "substantial equivalence" rationalization as an excuse to avoid any effective testing of GE foods to determine their safety. Such testing might seriously delay or even prevent the introduction of GE crops into the marketplace.

From the time in the early 1970s when advances in molecular biology led to the development of the techniques we call genetic engineering, until the mid 1990s, the organisms produced by genetic engineering were nearly all confined to laboratories or controlled factory settings. During this time there were almost no releases of genetically engineered organisms to the environment, as genetic engineering was used in basic research and to produce medically useful substances such as insulin. The unrestrained expansion of genetic engineering into agriculture during the past four years changed all that. By 1999 almost 80 million acres of North American farmland were planted with genetically engineered seed. This means massive releases of GEOs to the environment are now taking place. Genetic engineering now poses a very grave threat to the natural environment.

Historic Turning Point

We are now at a turning point in history. We can continue to allow the virtually unrestricted release of genetically engineered organisms to the environment, or we can bring this technology under strict control.

If we continue on our present path of unrestricted releases of GEOs, we will eventually live in a genetically engineered world, as the genome of each species now on earth is either deliberately altered by genetic engineering or indirectly altered by inheritance of transgenes from a genetically engineered organism. In such a world there would be nothing left of living nature, as every species would have been deprived of its genetic integrity, and every ecosystem would thereby have been irreversibly disrupted.

Of special concern to environmentalists should be the fact that trees are now being genetically engineered, and that it is proposed that entire forests be planted with these trees. One such proposal is for trees which produce no seeds, but divert the energy from seed production to more rapid growth of wood. A forest of such trees would wreak havoc on the food chain. Other GE trees that do produce seeds could cross with native varieties and damage forest ecosystems. Engineered trees which produced pollen (as might happen despite scientists' attempts to create sterile subspecies) could cross with native varieties miles away and damage forest ecosystems.

Fish, as well as other animals, are also being genetically engineered to grow more rapidly. If they are released to the environment (fish culture tanks often discharge during storm conditions), they may out-compete native species and thereby disrupt ecosystems.

There is evidence that soil organisms may be adversely affected bygenetically engineered crops. The Bt corn plant is engineered to contain a bacterial gene that causes production of an insecticide in every cell of the corn plant, including the edible corn ear and the roots. This toxin has been found to persist in the soil for months.

The promoters of genetic engineering show no sign that they are willing or able to impose limits on their applications of this new technology. It will therefore be left to the institutions of civil society— governments working with non-governmental organizations representing the concerned public (such as Sierra Club)—to set limits to how much further genetic engineering will be allowed to alter the earth's species.

Medical Uses

Promoters of the use of genetic engineering outside the laboratory claim that a moratorium or other controls on the planting of genetically engineered organisms as agricultural crops would mean an end to the uses of genetic engineering in the production of medically useful products. This is untrue. As long as proper precautions are taken to assure that the genetically engineered microorganisms used in production of pharmaceuticals or in scientific experiments are not released to the environment, such uses need not be prohibited. However, all applications of genetic engineering, including medical uses, carry some risk. Medical applications of genetic engineering should be approached with caution and not rushed to market. We believe that simpler, more traditional strategies for problem solving should always be considered when evaluating the production of transgenic organisms. This is especially relevant with respect to agricultural applications, as will be discussed in the topic below.

Feeding the World's Hungry?

Medical uses of genetic engineering may be prudent, but agricultural applications of GE are not. Yet the argument is being made by the biotech industry that if the genetic engineering of farm crops is not allowed to proceed, the poor people of the world will starve.

In fact there is more than enough food produced by conventional agriculture, without genetic engineering, to feed all of the world's people. One cause of hunger is the ineffective distribution of food. Genetic engineering may actually lead to more food insecurity and hunger because in poor countries it will lead to the planting of monoculture crops, highly vulnerable to disease and pests, in the place of resilient, diverse range of crops, and it will make farmers dependent on corporations that will demand payment for basic inputs such as seed, chemicals, and fertilizers.

Terms of trade between developed and less developed nations have often resulted in the best land in the poor countries being used to grow cash crops for export rather than food for consumption at home. Issues of equity and fairness have not been addressed by trade agreements. Certainly these problems call out for redress, but their solution isn't to increase the monopoly power of "life science" companies in the richest nations.

As Indian writer and activist Vandana Shiva summarized, "Millions of farmers in third world countries want to breed and grow the crop varieties that adapt to their diverse ecosystems. Plant biodiversity is essential for a balanced diet. Yet numerous crops are pushed to extinction with the introduction of GE crops."

Terminator Technology

Any claim by the corporations promoting agricultural biotechnology that they have the interests of the world's poor people at heart are refuted by the facts in the case of Terminator seed technology. This technology would protect the intellectual property interests of the seed company by making the seeds from a genetically engineered crop plant sterile, unable to germinate. Terminator would make it impossible for farmers to save seed from a crop for planting the next year, and would force them to buy seed from the supplier. In the third world, this inability to save seed could be a major, perhaps fatal, burden on poor farmers. The Sierra Club's Genetic Engineering Committee (GEC) believes that Terminator is a tool by which seed companies are trying to engineer their monopoly power into the genetic code.

Adding insult to injury in the Terminator technology story is the fact that our own tax dollars were used to develop Terminator. The U.S. Dept. of Agriculture played a major role in the development of Terminator technology. The USDA is actually part owner of the patent on this immoral technology.

The Genetic Engineering Committee

The Sierra Club's Genetic Engineering Committee (GEC) was formed in May 1999 to explore ways to mobilize the strength of Sierra Club, the largest grassroots environmental organization in the U.S., for the work of public education and regulatory reform that will be necessary to protect the natural environment and human health from the threats posed by the release of genetically engineered organisms.

An Educational Challenge

Genetic Engineering Committee members have found that the need for public education is great. Because of inadequate reporting by the U.S. media, many otherwise well educated people simply have not been told what genetic engineering is. We hear statements like, "If there is a moratorium on planting genetically engineered crops, doesn't that mean that no crops at all will be planted?" And, "Aren't all farm crops these days genetically engineered?"

The supporters of genetic engineering gladly fill this information vacuum with false statements. They claim that the selective breeding of plants and animals that has been done for centuries is genetic engineering. Supporters claim that modern genetic engineering is nothing more than an improved, more precise, high-tech form of conventional plant and animal breeding. Michael Khoo, in a letter published last year in the Toronto Globe and Mail, called this claim "... biotechnology's public-relations line that genetic engineering is no different from traditional breeding." His letter continued, "A potato can cross with a different strain of potato but, in 10 million years of evolution, it has never crossed with a chicken. Genetic engineering shatters these natural species boundaries, with completely unpredictable results. As a result of these risks, the British Medical Association has recently called for an open-ended moratorium on GE planting." Gene transfers occur in conventional breeding, but these transfers can only take place between individuals of the same species, or, in the case of hybridization, between individuals of closely related species. This is because conventional breeding relies on the normal reproductive processes of the plants or animals. Plants can be conventionally bred only with plants of the same species or, to make a hybrid, with closely related species. And animals can only be bred with other animals of the same or, in some instances, closely related species.

Genetic engineering is not bound by these limits in the possible exchanges of genes that can be made to occur using its techniques, which include the use of viruses as "vectors" to move foreign genes into host organisms. By means of genetic engineering, genes can be transferred from a plant to an animal; from an animal to a plant; from a bacteria to a plant, and between numerous other combinations of donor and recipient organisms. There have even been attempts to put human genes in plants and animals that are used as human food.

Why is this Important?

The changes caused by genetic engineering can be inherited by subsequent generations of the affected organism, and, once released to the environment, these organisms cannot be recalled—they will continue to pass on their spliced-in genes, or transgenes, to future generations. Many of the gene changes may turn out to have unexpected secondary effects. Serious errors in judgement might prove unrecallable as trillions of copies are broadcast via pollen and seed. Wild relatives of crops will also be affected, with possibly profound effects on the environment. For instance, genetically engineered cereals may cross with various grasses. Once this process begins, it will be for all practical purposes uncontrollable and unpredictable.

Biodiversity and Endangered Species

As environmentalists, one of our most basic concerns is the preservation of species. We live in a time when the rate of species extinction has increased drastically, primarily as a result of human activities. Now a new form of human activity, genetic engineering, may pose the ultimate threat to the survival of all species.

Many of those who are promoting genetic engineering give every indication that they regard life as a form of information technology: that genes are mere bundles of information to be transferred from one species to another on the basis of expediency and potential corporate cash-flow; that the natural barriers to genetic transfer that protect the integrity of species are mere inconveniences to be overcome; and that the very concept of species is an anachronism which it is now time to discard.

Because these principles are being put into application—genetically engineered organisms are now being made and released to the environment—we have to conclude that genetic engineering threatens the continued existence of all species as life-forms that are distinct from one another.

Genetic engineering should be considered an environmentally dangerous technology that is breaking down the barriers that have protected the integrity of species for millions of years. There are probably good reasons why it is impossible for a conventional plant breeder to combine plant genes with animal genes. Those reasons have to do with the very survival of life on earth, and we ignore them at our peril.

Another threat to biodiversity from genetic engineering is from toxins produced by GE crops. The finding last May that Bt corn pollen might be a threat to monarch butterflies provides an example. Genetically engineered Bt crops have the gene spliced-in from the Bt bacteria that codes for the production of the toxin that kills insect larvae. A Cornell study showed that this toxin kills the larvae of certain species of moths and butterflies. Other studies have indicated reduced viability of other nontarget beneficial insects, such as ladybugs and lacewings. Bt toxin also persists in the roots of the crops and in plant residues for a considerable time after the crop is harvested, which may have major adverse consequences for the millions of soil organisms that help maintain soil fertility.

Yet another threat to biodiversity is the out-crossing of herbicide resistance traits to native plants. There is already evidence of "superweeds" created by the spread of pollen carrying the herbicide resistance trait.

A threat to Organic Farming

The standards established by organic farmers categorically exclude genetically engineered crops from the organic food system. A problem arises from pollen drift from fields of GE crops planted too close to organic crops. The organic plants may become crossed with the GE plants and thereby contaminated with the spliced-in gene (transgene) from the GE crop. Then the crop grown next season from seed saved from what was an organic field will no longer be acceptable as organic—it will contain the transgene and will have to be considered genetically engineered. And in the case of crops in which the harvested portion of the plant is the seed, the presence of a transgene will immediately, in the first generation, make the crop not acceptable as organic. This problem of outflow of transgenes to organic crops is considered by organic growers to be very serious.

Another negative impact on organic farming is the expected resistance that insect pests will develop to Bt toxin. Organic farmers have been using Bt bacteria applied to crops in a spray as an organic method of controlling damaging insects. But genetically engineered Bt crops have the gene that codes for Bt toxin production spliced-in. By applying Bt bacterial sprays only occasionally, and because of the naturally limited quantity of the toxin present in the bacteria, organic farmers have avoided pest resistance problems. Now, with massive quantities of Bt toxin present in fields throughout the growing season, most of the insects susceptible to the toxin will be killed off, leaving a proportionately greater number of resistant insects alive. These Bt-resistant survivors will pass resistance traits into future generations. It is expected that resistance problems caused by genetically engineered Bt crops will render Bt sprays useless to organic farmers within a few years.

Health Issues

While Sierra Club is an environmental organization, we are concerned also with potential human health impacts of this new technology. Among the issues are the possible spread of allergens, the invitation which herbicide tolerant crops give to overuse of herbicides, possible adverse effects of new toxins (such as the Bt endotoxin) on some people, and the emergence of antibiotic resistance which may be fostered by the use of antibiotic resistance genes in almost all transgenic crops. New genes also alter the expression of native genes and so may change the nutritional benefits of foods and may also result in the overproduction of previously low-level natural toxins which exist in most foods. Health risks add to the environmental reasons for exercising caution.

The Precautionary Principle

The Genetic Engineering Committee strongly supports application of the precautionary principle to biotechnology issues and recognizes the limits inherent in present systems of risk assessment. Here is cogent statement of the precautionary principle from the Wingspread Consensus Statement on the Precautionary Principle, Jan, 1998: "When an activity raises threats of harm to the environment or human health, precautionary measures should be taken even if some cause and effect relationships are not fully established scientifically."

The participants at the conference said the following about risk assessment: "We believe existing environmental regulations and other decisions, particularly those based on risk assessment, have failed to protect adequately human health and the environment, the larger system of which humans are but a part."

Carolyn Raffensperger offered further commentary on risk assessment: "Participants [at the Wingspread conference] noted that current policies such as risk assessment and cost-benefit analysis give the benefit of the doubt to new products and technologies, which may later prove harmful. And when damage occurs, victims and their advocates have the difficult task of proving that a product or activity was responsible."

The precautionary principle is of the greatest importance when the damage from a new technology would be irreversible. This is the case with genetic engineering. Once they are released into the environment, genetically engineered organisms cannot be recalled. The Genetic Engineering Committee believes that genetically engineered farm crops are wrongly given the benefit of the doubt in the regulatory process, and that, under the precautionary principle, they should not be released into the environment or allowed to be part of the food supply.

The Regulatory Process

The federal government decided early in the development of genetically engineered crops that this was a technology where U.S. producers had an advantage which could be used to help them compete successfully in world markets. It was decided during the first Bush administration that the regulatory process for approval of GE crops would be streamlined. The Clinton administration continued this policy, with both President Clinton and Vice President Gore being strong supporters of agricultural biotechnology.

The regulatory inadequacies in the case of Bt potatoes are illustrative. The U.S. Food and Drug Administration (FDA) does not test the toxin in Bt potatoes for safety as a food additive because the toxin is a pesticide and therefore the U.S. Environmental Protection Agency (EPA) has the responsibility to assure its safety.

But the EPA tests only the Bt toxin, not the potatoes containing the toxin. So no one tests Bt potatoes for their safety as food, yet they become part of our food supply.

The FDA does not require labelling of Bt foods, because the agency is prohibited from requiring any information about pesticides on food labels, and because they consider GE foods to be substantially equivalent to conventional foods. Meanwhile, the U.S. Dept. of Agriculture pursues a role primarily of promotion of genetic engineering in agriculture, spending only a tiny fraction of its budget on safety testing of biotech foods.

As for testing GE crops for environmental hazards, there has been no environmental impact statement (EIS) done for a release of any genetically engineered crop. This is in violation of the National Environmental Policy Act (NEPA), which the EPA has the responsibility to administer.

Proposed Legislation

Laws are needed to require safety testing and labelling of GE crops. Also needed are mandatory environmental impact statements for every ecosystem into which any new GEO is to be introduced, and when applicable, involvement with the U.S. Fish and Wildlife Service. Liability issues also need to be addressed: clarification is needed as to who is responsible for the downstream effects of a company's product, including damage to organic producers and damage to the environment. Funding for agricultural research and development should be directed towards sustainable methods, rather than methods that perpetuate dependence on the chemical treadmill and agricultural biotechnology.

Definitions of Key Terms:

Biotechnology : A term now widely used to mean genetic engineering. In a larger sense, biotechnology is any use of biological processes to produce a desired result. Thus, the use of yeast to bake bread is a form of biotechnology which is not genetic engineering and which has been in use for centuries.

Genome : The complete set of genes of an individual organism, or the complete set of genes of all the individuals of a species.

Genetic engineering (GE) : The transfer of genes from one organism to another organism in ways that are not possible using conventional breeding methods. Genetic engineering bypasses the reproductive barriers that prevent genetic transfers between unrelated species, thus allowing transfer of genes from an organism of one species to another, completely unrelated species. Genetic engineering also includes methods of gene deletion and gene manipulation that are not possible using conventional breeding methods.

Genetically engineered organism (GEO) : Any living thing that has had its genetic structure altered by genetic engineering. A genetically engineered organism is also called a genetically modified organism (GMO), a genetically altered organism, or in certain cases, a transgenic organism.

Recombinant DNA technology : The technique, also called gene splicing, that made possible the first application of genetic engineering, in 1973. A section of DNA molecule which constitutes a gene, the basic unit that determines an inherited trait, is cut from the molecule and spliced into another DNA molecule in another organism. The two organisms need not be of the same species or even closely related. Thus, using recombinant DNA techniques, genes from bacteria have been spliced into corn plants and DNA from a fish has been spliced into strawberry plants. It is also possible to splice plant DNA into an animal.

Transgene : A gene from one organism transferred into another (usually unrelated) organism by means of genetic engineering.

Facts on Animal Cloning

- Animals bred through cloning are born to mothers in the usual way and grow up just likeother animals.
- Cloning is simply a breeding technique that creates an identical twin of an existing animal.
- Livestock cloning does not change the animal in any way. It is not genetic engineering. There is no such thing as "cloned food." Animal clones will be used as breeding stock.
- Animal cloning offers great benefits to consumers, farmers, and endangered species, including:
- Cloning enhances the availability of the best possible livestock by allowing farmers to be certain of the genetic makeup of a

particular animal, thus allowing them to better produce high-quality, safe, and healthy food.

- Cloning can offer a tremendous advantage for farmers whose livelihoods depend on selling high-quality meat and dairy products. The breeding technique allows a greater number of farmers the ability to preserve and extend proven, superior genetics. Ranchers would be able to select and propagate the best animals — beef cattle that have lean but tender meat, and are disease-resistant.

- Cloning reproduces the strongest, healthiest animals, thus optimizing animal wellbeing and may minimize the need for veterinary intervention.

- Cloning can be used to protect endangered species. For example, in China, panda cells are kept on reserve should the panda's numbers be threatened by extinction.

- In January 2008, the U.S. Food and Drug Administration (FDA) published a risk assessment that concluded that meat and milk products from cloned animals and their offspring are safe for human consumption, and no different from foods produced through other breeding methods.

- Under current FDA labelling guidelines, food products from animal clones will not require special labelling because these foods have been deemed to be nutritionally and compositionally equivalent to products from conventionally-bred animals.

- Animal clones will primarily be used as breeding stock to improve the health and quality of animals used for food production. So, most consumers will likely never eat an animalclone; rather, meat and milk products in the marketplace will come from the offspring of animal clones. These offspring would be bred through other conventional breeding techniques, and not be clones themselves.

Modern Biotechnology and Development

This study was commissioned by the World Health Organization (WHO) to establish a knowledge base for evaluating the application of modern biotechnology in food production. The study does not seek to address all issues and evidence in detail, but rather aims to place in context the overall impact of this technology on human health and development. The study reviews evidence in several broad areas related to the use of genetically modified (GM) organisms in the food supply (GM foods), including a review of GM food products currently available, the assessment of risks and benefits, the broader impact on societies, and the existing regulatory capacity in countries. The evidence was collected and collated by WHO with the support of a background group of external experts (list of experts-annex 1). Data for the study were gathered through traditional methodology as well as through an open questionnaire and an Internet-based electronic discussion process. Preliminary results were discussed at a broad stakeholder meeting held in 2003 (list of participants-annex 1), informing further data search and revision.

The first GM food (delayed-ripening tomato) was introduced on the US market in the mid-1990s. Since then, GM strains of maize, soybean, rape and cotton have been adopted by a number of countries and marketed internationally. In addition, GM varieties of papaya, potato, rice, squash and sugar beet have been trialed or released. It is estimated that GM crops cover almost 4% of total global arable land.

The development of GM organisms (GMOs) offers the potential for increased agricultural productivity or improved nutritional value that can contribute directly to enhancing human health and development. From a health perspective, there may also be indirect benefits, such as reduced agricultural chemical usage and enhanced farm income, and improved crop sustainability and food security, particularly in developing countries. Contradictory findings for such benefits sometimes reflect different regional or agricultural conditions.

The use of GMOs may also involve potential risks for human health and development. Many genes used in GMOs have not been in the food supply before. While new types of conventional food crops are not usually subject to safety assessment before marketing, assessments of GM foods were undertaken before the first crops were commercialized. To provide international consistency in the assessment of GM foods, principles developed by the Codex Alimentarius Commission (a joint programme of WHO and the Food and Agriculture Organization of the United Nations; FAO) now cover food safety, while the Cartagena Protocol on Biosafety covers environmental safety of GMOs. Many countries have established specific premarket regulatory systems in accordance with this international guidance that require a case-by-case risk assessment of each GM food. Risk assessment methodology undergoes continuous improvements, a fact that is recognized by the Codex principles, including the need for risk assessments to consider both the intended and unintended effects of such foods in the food supply. GM foods currently traded on the international market have passed risk assessments in several countries and are not likely, nor have been shown, to present risks for human health.

Although risk-assessment systems have been in use for some time, the perception of GM food among consumers has not always recognized these assessments. One explanation is that many national food-safety systems have had problems performing good risk communication in this area. In many countries, social and ethical considerations may cause also resistance to modifications which interfere with genes. These conflicts often reflect deeper issues related to the interaction of human society with nature — issues that should be taken seriously in any communication effort. However, while in many regions, food is clearly considered part of historical identity and societal life, scepticism towards GM food is not necessarily linked to traditionalism or to absence of knowledge about this new technology. Investigations of public perception indicate that the sceptical consumer will acknowledge arguments both for and against GM food and, in general, does not demand 'zero risk'. Likewise, it has been seen that critical attitudes towards GM food are not necessarily linked to a negative attitude towards the use of biotechnology as such, as demonstrated by a generally positive attitude towards the use of biotechnology in modern medicine. The issue of benefit to society therefore seems to constitute an important aspect related to acceptance of new technology.

Intellectual property rights are an important part of the GM food debate. Problems of assuring equal access to genetic resources, sharing benefits on a global level, and avoiding monopolization exist for GM food as for other uses of gene technology. Related to this are concerns about a growing influence of the chemical industry in seed markets. Sustainable agriculture and biodiversity are likely to benefit most when a rich variety of crops are planted, and a potential exclusive use of certain chemical-resistant GM crops could be seen to create dependency.

Conflicting assessments and incomplete substantiation of the benefits, risks and limitations of GM food have added to existing controversies. During a famine situation in southern Africa in 2002, the reluctance among several recipient countries to receive GM food aid was not primarily linked to health or environment issues, but to socioeconomic, ownership and ethical issues. Such controversies have not only highlighted the wide range of opinions within and between Member States, but also the existing diversity in regulatory frameworks and principles for assessing the benefits and risks of GM food. In addition, many developing countries cannot afford to build the separate capacities required for effective regulation of GM foods, which again underlines the benefits that could be derived from international work for broader evaluations of GM food applications.

At the international level, 15 legally binding instruments and non-binding codes of practice address some aspect of GMO regulation or trade. Such sector-based regulations increase the already overstretched capacity of developing countries, and present challenges to develop a fully coherent policy and regulatory framework for modern biotechnology. This study makes the case for the need for an evidence base to facilitate a more coherent evaluation of the application of modern food biotechnology and the use of GM foods. Such an evidence base should: deal with the assessment of human health and environmental risk as well as benefit; evaluate socioeconomic factors, including intellectual property rights; and consider ethical aspects. International harmonization in all these areas is a prerequisite for the prudent, safe and sustainable development of any new technology, including the use of biotechnology to produce food. Work towards such harmonization can only move forward through inter-sectoral collaboration and would therefore necessarily extend beyond the WHO mandate into the mandates of several other international organizations. This report should be seen as one possible starting point for further inter-sectoral discussions.

Goals and Terms of Reference

The World Health Organization (WHO) commissioned this study to establish a broad knowledge base for Member States, international standard-setting bodies and other stakeholders, in order to achieve transparent and inclusive consensus on the evaluation and application of modern biotechnology in the production of food. The aim of this study is to determine the significance of the application of modern biotechnology to food production in terms of human health and development. The study does not seek to address all issues and evidence in detail, but rather to place in context the overall impact that modern food biotechnology may have on human health and development. It is intended to serve as a scientific basis for potential discussion by the governing bodies of WHO.

The study reviews evidence in five broad areas:

1. Current use, research and impending development of foods produced through modern biotechnology, and their significance for human health and development.

2. Risk assessments of present and future products of modern biotechnology in relation to food safety, human nutrition and environmental health.

3. The significance of modern food biotechnology for food security, and the impact of intellectual property rights on research.

4. National capacity for risk assessment and management.

5. The impact of modern food biotechnology on civil society, considering social and ethical concerns.

Methodology

A background group consisting of experts from various Member States (*Annex 1*) established the terms of reference of the study and a guidance document that directed a small team within WHO to gather the evidence. Members of the background group also assisted in data gathering.

Data were gathered using extensive literature and Internet searches, and through a questionnaire supported by approximately 120 responses which was circulated to a broad range of stakeholders in May 2002. The comments received from an electronic stakeholder discussion held between January and April 2003 have also been incorporated. The opinions of participants who attended a stakeholder meeting on 5–6 June 2003 in Geneva, comprising representatives from governments, consumers, industry, research and non-governmental organizations (NGOs), from developed and developing countries, have also been included.

The focus on including a broad basis of scientific evidence as well as descriptions of opinions from a broad group of stakeholders has resulted in a list of references which includes documentation from many Internet sites. Documentation originating solely from Internet sites should not, in general, be treated or presented as documentation derived from peer-reviewed literature; however, it has been considered necessary in this study to include data and information presented from both sources, with a clear indication of when information is available solely from Internet sources.

This study focuses on the application of modern biotechnology (especially recombinant DNA technology) to organisms used to produce food.

The application of modern biotechnology to food production presents new opportunities and challenges for human health and development. Recombinant gene technology, the most well-known modern biotechnology, enables plants, animals and microorganisms to be genetically modified (GM) with novel traits beyond what is possible through traditional breeding and selection technologies. It is recognized that techniques such as cloning, tissue culture and marker-assisted breeding are often regarded as modern biotechnologies, in addition to genetic modification.

The inclusion of novel traits potentially offers increased agricultural productivity, or improved quality and nutritional and processing characteristics, which can contribute directly to enhancing human health and development. From a health perspective, there may also be indirect benefits, such as reduction in agricultural chemical usage, and enhanced farm income, crop sustainability and food security, particularly in developing countries.

The novel traits in genetically modified organisms (GMOs) may also, however, carry potential direct risks to human health and development. Many, but not all, genes and traits used in agricultural GMOs are novel and have no history of safe food use. Several countries have instituted guidelines or legislation for mandatory premarket risk assessment of GM food. At the international level, agreements and standards are available to address these concerns.

GMOs may also affect human health indirectly through detrimental impacts on the environment, or through unfavourable impacts on economic (including trade), social and ethical factors.

These impacts need to be assessed in relation to the benefits and risks that may also arise from foods that have not been genetically modified. For example, new, conventionally bred varieties of a crop plant may also have impacts — both positive and negative — on human health and the environment.

Recent International Controversies and Study Initiative

Conflicting assessments and incomplete substantiation of the benefits, risks and limitations of GM food organisms by various scientific, commercial, consumer and public organizations have resulted in national and international controversy regarding their safe use as food and safe release into the environment. An example is the debate on food aid that contained GM material offered to countries in southern Africa in 2002, after 13 million people faced famine following failed harvests. This international debate highlighted several important issues, such as health, safety, development, ownership and international trade in GMOs. Such controversies have not only highlighted the wide range of opinions within and between Member States, but also the existing diversity in regulatory frameworks and principles for assessing benefits and risks of GMOs. In view of this lack of consensus, the Fifty-third World Health Assembly in 2000 adopted resolution WHA53.15 (WHO 2000b), according to which WHO should strengthen its capacity to support Member States to establish the scientific basis for decisions on GM food organisms, and ensure the transparency, excellence and independence of opinions delivered. This study aims to provide an evidence base to assist individual Member States in their consideration of the application of modern food biotechnology and the use of GM foods, and to facilitate greater international harmonization in this regard.

Current Use, Research and Impending Development of Foods Produced through Modern Biotechnology

Foods produced through modern biotechnology can be categorized as follows:

1. Foods consisting of or containing living/viable organisms, e.g. maize.

2. Foods derived from or containing ingredients derived from GMOs, e.g. flour, food protein products, or oil from GM soybeans.

3. Foods containing single ingredients or additives produced by GM microorganisms (GMMs), e.g. colours, vitamins and essential amino acids.

4. Foods containing ingredients processed by enzymes produced through GMMs, e.g. high-fructose corn syrup produced from starch, using the enzyme glucose isomerase (product of a GMM).

This study, however, makes no attempt to discriminate between the various categories, and the discussion that follows describes the current and future applications of modern biotechnology in the production of crops, livestock, fish and microorganisms in food production.

Crop Breeding and the Introduction of GM crops for food Production

Conventional breeding, especially of crops, livestock and fish, focuses principally on increased productivity, increased resistance to diseases and pests, and enhanced quality with respect to nutrition and food processing. Advances in cellular genetics and cell biology methods in the 1960s contributed to the so-called 'green revolution' that significantly increased varieties of staple food crops containing traits for higher yield and resistance to diseases and pests in a number of both developed and developing countries (Borlaug 2000). A key driver of the green revolution was to improve the potential to provide sufficient food for all. The intensification and expansion of agriculture brought about by these methods and agricultural systems have, however, also resulted in new forms of health and environmental risks through, for example, increased use of agrochemicals and intensified cultivation resulting in soil erosion.

The development of molecular biology in the 1970s and 1980s introduced more direct methods for the analysis of genetic sequences and allowed the identification of genetic markers for desired traits. Such marker-assisted breeding methods are the basis of some current conventional breeding strategies. Whereas modern methods of breeding have significantly increased crop yields over the past 50 years, the future potential of these methods is constrained by the limitations in the natural diversity of trait genotype within crop species and sexual-compatibility boundaries between crop types.

To overcome these problems, a number of interested groups (scientists, farmers, governments, agricultural companies) have since the 1980s considered other means to achieve the objectives of improved yields, sustainable agricultural systems, and improvements in human and animal health and the environment. This includes the use of more modern methods to introduce novel traits, such as tolerance to drought, salt, or pests. To achieve these objectives, various public and, more recently, private research programmes have aimed to improve the understanding of and links between crop performance and molecular genetics. With the development and use of recombinant DNA in the 1980s, a tool to overcome the limitation of species incompatibility was found. Modern biotechnology employs molecular techniques to identify, select and modify DNA sequences for a specific genetic trait (e.g. insect resistance) from a donor organism (microorganism, plant or animal), and transfer the sequence to the recipient organism so that it expresses this trait.

Various transformation methods are used to transfer recombinant DNA into recipient species to produce a GMO. For plants, these include transformation mediated by *Agrobacterium tumefaciens* (a common soil bacterium that contains genetic elements for infection of plants) and biolistics — shooting recombinant DNA placed on microparticles into recipient cells. The methods used in the transformation of various animal species include microinjection, electroporation and germ-line cells (FAO/WHO 2003a). The success rate of transformations in animals tends to be lower than in plants, and to vary from species to species, thus requiring the use of many animals.

Genetic modification is often faster than conventional breeding techniques, as stable expression of a trait is achieved using far fewer breeding generations. It also allows a more precise alteration of an organism than conventional methods of breeding, as it enables the selection and transfer of a specific gene of interest. However, with the present technology, in many cases it leads to random insertion in the host genome, and consequently may have unintended developmental or physiological effects. However, such effects can also occur in conventional breeding and the selection process used in modern biotechnology aims to eliminate such unintended effects to establish a stable and beneficial trait.

It should be noted that conventional breeding programmes directed by the molecular analysis of genetic markers are also of critical importance to modern plant and animal breeding. However, human and environmental health consequences of these techniques are not considered here.

GM Crops Currently in Commercial Production

At present, only a few GM crops are permitted for food use and traded on the international food and feed markets. These include herbicide-and insect-resistant maize (*Bt*[1] maize), herbicide-resistant soybean, rape (canola) oilseed, and insect-and herbicide-resistant cotton (primarily a fibre crop, though refined cottonseed oil is used as food). In addition, several government authorities have approved varieties of papaya, potato, rice, squash, sugar beet and tomato for food use and environmental release. The latter crops, however, are currently grown and traded only in a limited number of countries, mainly for domestic consumption.

The regulatory status of GM crops varies among the countries that permit their use and updates can be found on various web-sites, including those of the Organisation for Economic Co-operation and Development (OECD) and the International Centre for Genetic Engineering and Biotechnology (ICGEB).

Future Trends in GM Crops

The commercial introduction of transgenic crop plants with agronomic traits is often referred to as the first generation of transgenic plants. Further development of GM crops with agronomic traits is continuing, and production of a range of GM crops with enhanced nutritional profiles is also under way (PIFB 2001). Various novel traits are currently being tested in laboratories and field tests in a number of countries. Many of these second-generation GM crops are still in the development stage and are unlikely to enter the market for several years.

The key areas of research and development (R&D) in plants are (i) agronomic traits and (ii) altered nutrition and composition.

PPAgronomic Traits

Pest and disease resistance: In the short term, most newly commercialized GM crops will continue to concentrate on agronomic traits, especially herbicide resistance and insect resistance and, indirectly, yield potential (PIFB 2001). R&D in this area aims to:

- introduce herbicide-resistance traits in a broader range of varieties of maize, soybean and canola;
- broaden the range of herbicides that can be used in combination with the transgenic herbicide-resistant crop, such as introduction of tolerance to the herbicides bromoxynil, oxynil and sulfonylurea; and
- stack novel genes for insect resistance in plants, such as novel *Bt* variants containing different toxins.

Virus resistance: Virus resistance could be extremely important to improving agricultural productivity (Thompson 2003). Field tests of the following virus-resistant crops are currently being conducted in various parts of the world: sweet potato (feathery mottle virus); maize (maize streak virus); and African cassava (mosaic virus). These crops may be available for commercialization within the next 3–5 years. Because of its complex genome, work on wheat resistant to the barley yellow-dwarf virus has made little progress and is still undergoing laboratory investigation. Resistance to nematodes (root worms) in a GM potato has also been achieved.

Altered Nutrition and Composition

Vitamin-A-enhanced rice: The best-known example of a GM crop conferring enhanced nutritional properties is rice containing a high level of beta-carotene — a vitamin A precursor (so-called 'golden rice') (Potrykus 2000). Vitamin A is essential for increasing resistance to disease, protecting against visual impairment and blindness, and improving the chances of growth and development. Vitamin A deficiency (WHO/UNICEF 1995) is a public health problem that contributes to severe illness and childhood mortality. This preventable condition increases the burden of disease on the health systems of developing countries. A number of strategies have been suggested for combating vitamin A deficiency, including dietary approaches (e.g. fortification of foods) and supplementation via pills (WHO 2000c). Within the context of improving the supply of vitamin A, the usefulness of vitamin A-enhanced rice has been discussed in various forums, such as an electronic forum coordinated by the Food and Agriculture Organization of the United Nations (FAO) in 2000 (FAO 2000).

Vitamin A-enhanced rice and maize varieties are at present being developed for cultivation in developing countries. Current efforts are aimed at ensuring that vitamin A in rice can be absorbed efficiently in the human gut. Once this is resolved, 300 grams of transgenic rice could make a significant contribution to the daily human requirement for vitamin A.

'High iron' rice : Prevalence of iron deficiency is very high in those parts of the world in which rice is the daily food staple (WHO 2000a). This is because rice has a very low iron content. Transgenic rice seeds with the iron-carrier protein ferritin from soy were found to contain twice as much iron as seeds of non-transformed rice (Gura 1999). Rice has been transformed with three genes which increase iron storage in rice kernels and iron absorption from the digestive tract (Lucca et al. 2002).

Improved protein content : Researchers are also investigating methods that could improve the protein content of staple vegetables, such as cassava, plantain and potato (PIFB 2001). Results from greenhouse trials show that these tubers have 35–45% more protein, and enhanced levels of essential amino acids.

Removing allergens and antinutrients: Cassava roots naturally contain high levels of cyanide. As they are a staple food in tropical Africa, this has led to high blood-cyanide levels which have harmful effects. Application of modern biotechnology to decrease the levels of this toxic chemical in cassava would reduce its preparation time. In potatoes, insertion of an invertase gene from yeast reduces the natural levels of glycoalkaloid toxin (Buchanan et al. 1997).

The allergenic protein in rice has been reduced by modifying its biosynthetic pathway (PIFB 2001). The significance in human allergenicity of these lower levels has not been demonstrated. There is also work to reduce allergenicity in wheat (Buchanan et al. 1997). This work involves inserting a thioredoxin-biosynthesis gene to break the disulfide bonds in the offending protein but without interfering with the functionality of the wheat proteins.

Altered starch and fatty acid profile: In the quest to provide healthier foods, there is an effort to increase the starch content of potatoes so that they absorb less fat during frying (PIFB 2001). To create healthier fats, the fatty-acid composition of soy and canola has been altered to produce oils with reduced levels of saturated fats. R&D is currently focusing on GM soybean, oilseed rape and oil palm (PIFB 2001). Two GM crops of this nature have been approved in the United States of America (USA) for growing and food/feed use — high oleic acid soy and high lauric acid oilseed rape (Agbios 2005). High oleic acid soy is also permitted as food in Australia and Canada. R&D is in the early stages with respect to oils with improved nutritional value.

Increased antioxidant content: The lycopene and lutein contents of tomatoes have been increased as have isoflavones in soy (WHO 2000c). These phytonutrients are known to improve health or prevent disease. Research in this area is at a relatively early stage of development, as knowledge of phytonutrients is limited and not all phytonutrients are beneficial.

Environmental stresses: Tolerance to environmental-stress factors through genetic modification is an area that is in the early stages of R&D (PIFB 2001). Resistance to salinity and drought are being researched intensively. Salinity is estimated to affect 20% of agricultural land and 40% of irrigated land worldwide. Salt and drought tolerance involve numerous genes interacting in a complex manner. Owing to this multigenic character, conventional breeding techniques have had little success in the generation of salt-or drought-tolerant varieties. Salt tolerance may be conferred to sensitive crops by the transfer of multiple genes linked to a relevant pathway from a tolerant crop. The likely time frame for commercialization for such GM crops is unknown.

Tolerance to aluminium (a growth-limiting factor in acid soils) is in the early phase of R&D for several crops, including papaya, tobacco, rice and maize, but they are not expected to be in commercial use for several years.

Attempts have been made to improve the photosynthetic system in plants through genetic modification. Crops such as maize and sugar cane are more efficient in converting energy into sugars than most broadleaf crop plants. By introducing genes for more efficient photosynthesis from one crop to another, efficiency could be improved by 10% with an enhancement in yield. The likely time frame of commercialization is unknown.

Male-sterility traits have been introduced for obtaining 100% hybrid sowing-seed for the purposes of environmental containment of GM crops. Various male-sterile maize varieties have been approved for market introduction in the USA. In addition, various male-sterile rapeseed and canola varieties have been approved for environmental release and food use in the European Union (EU), Canada and the USA. Another strategy for containing gene flow between plants attempts to introduce asexual seed propagation in crops (seed production without the need to pollinate). None of the above-mentioned strategies has proved applicable to all crop species, and a combination of approaches may prove most effective.

Livestock and Fish

In terms of food production, the application of modern biotechnology to livestock falls into two main areas: animal production and human nutrition. Many of the applications discussed below are in the early stages of R&D.

Fish

The projected increasing demand for fish suggests that GM fish may become important in both developed and developing countries. Enhanced-growth Atlantic salmon containing a growth hormone gene from Chinook salmon is likely to be the first GM animal on the food market (FAO/WHO 2003a). These fish grow 3–5 times faster than their non-transgenic counterparts, to reduce production time and increase food availability. At least eight other farmed fish species have been genetically modified for growth enhancement. Other fish in which genes for growth hormones have been experimentally introduced include grass carp, rainbow trout, tilapia and catfish (PIFB 2003; PIFB/FDA 2003). In all cases, the growth-hormone genes are of fish origin.

To address some of the practical problems of aquaculture, research attempts are seeking to improve disease resistance by producing Atlantic salmon with a rainbow trout lysozyme cDNA. Lysozyme has antimicrobial properties against fish pathogens such as *Vibrio*, *Aeromonas* and *Yersinia*. Another type of antimicrobial protein (silk moth cecropin) is under investigation in catfish (Dunham et al. 2002). This would improve catfish resistance to diseases such as enteric septicaemia.

The farming of carnivorous fish species, such as trout and salmon, has led to overfishing of sand eels and capelin. To tackle this problem, research is looking into the possibility of altering the metabolism of these species by improving their digestion of carbohydrates, to enable a shift to a more plant-based diet.

Lack of cold tolerance in warm-water species such as the common carp and tilapia can lead to significant stock losses in winter. The suggestion of work in this area is to alter the molecular conformation of lipids, thus increasing membrane fluidity. To extend the geographical range of fish farming, an antifreeze gene from one fish species is transferred to the species of interest. Although freeze-resistant strains of Atlantic salmon have been produced, the level of antifreeze protein secreted by the salmon was insufficient to have a significant impact on the freezing point of blood (Fletcher et al. 2002).

Livestock and Poultry

Foods derived from GM livestock and poultry are far from commercial use. Several growth-enhancing novel genes have been introduced into pigs that have also affected the quality of the meat, i.e. the meat is more lean and tender (FAO/WHO 2003a). This research was initiated over a decade ago, but owing to some morphological and physiological effects developed by the pigs, these have not been commercialized.

Many modifications to milk have been proposed that either add new proteins to milk or manipulate endogenous proteins (PIFB 2002b). Recently, researchers from New Zealand developed GM cows that produce milk with increased levels of casein protein. Use of such protein-rich milk would increase the efficiency of cheese production. Other work aims to reduce the lactose content of milk, with the intent of making milk available to the population of milk-intolerant individuals.

Other applications of genetic modification in animal production in the early stages of R&D include improvement of disease resistance, increased birth rates in sheep, altered sex ratio in poultry, increased egg production in poultry by creating two active ovaries, and improved feed conversion in the 'enviropig' (environmentally friendly pigs that excrete less phosphorus). Most of this work is still theoretical and therefore estimates of time frames for possible commercial introductions of any of these applications are unavailable.

Factory Farming

Factory farming is a term referring to the process of raising livestock in confinement at high stocking density, where a farm operates as a factory — a practice typical in industrial farming by agribusinesses. The main product of this industry is meat, milk and eggs for human consumption. However, there have been issues regarding whether factory farming is sustainable and ethical.

Confinement at high stocking density is one part of a systematic effort to produce the highest output at the lowest cost by relying on economies of scale, modern machinery, biotechnology, and global trade. Confinement at high stocking density requires antibiotics and pesticides to mitigate the spread of disease and pestilence exacerbated by these crowded living conditions. In addition, antibiotics are used to stimulate livestock growth by killing intestinal bacteria. There are differences in the way factory farming techniques are practiced around the world. There is a continuing debate over the benefits and risks of factory farming. The issues include the efficiency of food production; animal welfare; whether it is essential for feeding the growing global human population; the environmental impact and the health risks.

Agriculture adopted more intensive methods during the eighteenth century. With this growth in production best characterized by the Agricultural Revolution, where improvements in farming techniques allowed for significantly improved yields, and supported the urbanization of the population during the Industrial Revolution.

Innovations in agriculture beginning in the late nineteenth century paralleled developments in mass production in other industries. The identification of nitrogen and phosphorus as critical factors in plant growth led to the manufacture of synthetic fertilizers, making possible more intensive types of agriculture.

The first animals to be factory farmed were chickens. The discovery of vitamins and their role in animal nutrition, in the first two decades of the twentieth century, led to vitamin supplements, which allowed chickens to be raised indoors. The discovery of antibiotics and vaccines facilitated raising livestock in larger numbers by reducing disease. Chemicals developed for use in World War II gave rise to synthetic pesticides. Developments in shipping networks and technology have made long-distance distribution of agricultural produce feasible.

According to the BBC, factory farming in Britain began in 1947 when a new Agriculture Act granted subsidies to farmers to encourage greater output by introducing new technology, in order to reduce Britain's reliance on imported meat. The United Nations writes that intensification of animal production was seen as a way of providing food security.

In 1960s North America, pigs and cows began to be raised on factory farms. This practice then spread to Western Europe. In Britain, the agriculture correspondent of *The Guardian* wrote in 1964:

Factory farming is greatly debated throughout Australia, with several people disagreeing with the methods and ways in which the animals in factory farms are treated. Often animals are under stress from being kept in confined spaces, so will attack each other. This results in them having beaks, tails and teeth removed. Many piglets will die of shock after having their teeth and tails removed. This is due to the fact that painkilling medicines are not used in these operations. Others say that factory farms are a great way to gain space, with animals such as chickens being kept in spaces smaller than an A4 page. From its American and West European heartland factory farming became globalised in the later years of the twentieth century and is still expanding and replacing traditional practices of stock rearing in an increasing number of countries. In 1990 factory farming accounted for 30% of world meat production. By 2005 this had risen to 40%.

Nature of the Practice

Factory farms hold large numbers of animals, typically cows, pigs, turkeys, or chickens, often indoors, typically at high densities. The aim of the operation is to produce as much meat, eggs, or milk at the lowest possible cost. Food is supplied in place, and a wide variety of artificial methods are employed to maintain animal health and improve production, such as the use of antimicrobial agents, vitamin supplements, and growth hormones. Physical restraints are used to control movement or actions regarded as undesirable. Breeding programs are used to produce animals more suited to the confined conditions and able to provide a consistent food product.

The distinctive characteristic of factory farms is the intense concentration of livestock. At one farm (Farm 2105) run by Carrolls Foods of North Carolina, the second-largest pig producer in the U.S., twenty pigs are kept per pen and each confinement building or "hog parlor" holds 25 pens. The company's chief executive officer, F.J. "Sonny" Faison, has said: "It's all a supply-and-demand price question ... The meat business in this country is just about perfect, uncontrolled supply-and-demand free enterprise. And it continues to get more and more sophisticated, based on science. Only the least-cost producer survives in agriculture." Faison states:

The large concentration of animals, animal waste, and the potential for dead animals in a small space poses ethical issues. It is recognized that some techniques used to sustain intensive agriculture can be cruel to animals. As awareness of the problems of intensive techniques has grown, there have been some efforts by governments and industry to remove inappropriate techniques.

Farm Sanctuary defines factory farming as "an attitude that regards animals and the natural world merely as commodities to be exploited for profit." The group contends "this attitude has led to institutionalized animal cruelty, massive environmental destruction and resource depletion, and animal and human health risks."

In the UK, the Farm Animal Welfare Council was set up by the government to act as an independent advisor on animal welfare in 1979 and expresses its policy as five freedoms: from hunger & thirst; from discomfort; from pain, injury or disease; to express normal behaviour; from fear and distress.

There are differences around the world as to which practices are accepted and there continue to be changes in regulations with animal welfare being a strong driver for increased regulation. For example, the EU is bringing in further regulation to set maximum stocking densities for meat chickens by 2010, where the UK Animal Welfare Minister commented, "The welfare of meat chickens is a major concern to people throughout the European Union. This agreement sends a strong message to the rest of the world that we care about animal welfare."

However, given the assumption that intensive farming techniques are a necessity, it is recognized that some apparently cruel techniques are better than the alternative. For example, in the UK, de-beaking of chickens is deprecated, but it is recognized that it is a method of last resort, seen as better than allowing vicious fighting and ultimately cannibalism. Between 60 and 70 percent of six million breeding sows in the U.S. are confined during pregnancy, and for most of their adult lives, in 2 ft (0.61 m) by 7 ft (2.1 m) gestation crates. According to pork producers and many veterinarians, sows will fight if housed in pens. The largest pork producer in the U.S. said in January 2007 that it will phase out gestation crates by 2017. They are being phased out in the European Union, with a ban effective in 2013 after the fourth week of pregnancy. With the evolution of factory farming, there has been a growing awareness of the issues amongst the wider public, not least due to the efforts of animal rights and welfare campaigners. As a result gestation crates, one of the more contentious practices, are the subject of laws in the U.S., Europe and around the world to phase out their use as a result of pressure to adopt less confined practices'.

Human Health Impact

According to the U.S. Centres for Disease Control and Prevention (CDC), farms on which animals are intensively reared can cause adverse health reactions in farm workers. Workers may develop acute and chronic lung disease, musculoskeletal injuries, and may catch infections that transmit from animals to human beings (such as tuberculosis).

Pesticides are used to control organisms which are considered harmful and they save farmers money by preventing product losses to pests.

In the US, about a quarter of pesticides used are used in houses, yards, parks, golf courses, and swimming pools and about 70% are used in agriculture. However, pesticides can make their way into consumers' bodies which can cause health problems. One source of this is bioaccumulation in animals raised on factory farms.

The CDC writes that chemical, bacterial, and viral compounds from animal waste may travel in the soil and water. Residents near such farms report problems such as unpleasant smell, flies and adverse health effects.

The CDC has identified a number of pollutants associated with the discharge of animal waste into rivers and lakes, and into the air. The use of antibiotics may create antibiotic-resistant pathogens; parasites, bacteria, and viruses may be spread; ammonia, nitrogen, and phosphorus can reduce oxygen in surface waters and contaminate drinking water; pesticides and hormones may cause hormone-related changes in fish; animal feed and feathers may stunt the growth of desirable plants in surface waters and provide nutrients to disease-causing micro-organisms; trace elements such as arsenic and copper, which are harmful to human health, may contaminate surface waters.

In the European Union, growth hormones are banned on the basis that there is no way of determining a safe level. The UK has stated that in the event of the EU raising the ban at some future date, to comply with a precautionary approach, it would only consider the introduction of specific hormones, proven on a case by case basis. In 1998, the European Union banned feeding animals antibiotics that were found to be valuable for human health. Furthermore, in 2006 the European Union banned all drugs for livestock that were used for growth promotion purposes. As a result of these bans, the levels of antibiotic resistance in animal products and within the human population showed a decrease.

The various techniques of factory farming have been associated with a number of European incidents where public health has been threatened or large numbers of animals have had to be slaughtered to deal with disease. Where disease breaks out, it may spread more quickly, not only due to the concentrations of animals, but because modern approaches tend to distribute animals more widely. The international trade in animal products increases the risk of global transmission of virulent diseases such as swine fever, BSE, foot and mouth and bird flu.

Methicillin-resistant Staphylococcus aureus (MRSA) has been identified in pigs and humans raising concerns about the role of pigs as reservoirs of MRSA for human infection. One study found that 20% of pig farmers in the United States and Canada in 2007 harbored MRSA. A second study revealed that 81% of Dutch pig farms had pigs with MRSA and 39% of animals at slaughter carried the bug were all of the infections were resistant to tetracycline and many were resistant to other antimicrobials. A more recent study found that MRSA ST398 isolates were less susceptible to tiamulin, an antimicrobial used in agriculture, than other MRSA or methicillin susceptible *S. aureus.* Cases of MRSA have increased in livestock animals. CC398 is a new clone of MRSA that has emerged in animals and is found in intensively reared production animals (primarily pigs, but also cattle and poultry), where it can be transmitted to humans. Although being dangerous to humans CC398 is often asymptomatic in food-producing animals.

A 2011 study reported that according to a nationwide study nearly half of the meat and poultry sold in U.S. grocery stores—47 percent—was contaminated with S. aureus, and more than half of those bacteria—52 percent—were resistant to at least three classes of antibiotics. Although Staph should be killed with proper cooking, it may still pose a risk to consumers through improper food handling and cross-contamination in the kitchen. The senior author of the study said, "The fact that drug-resistant S. aureus was so prevalent, and likely came from the food animals themselves, is troubling, and demands attention to how antibiotics are used in food-animal production today." In April 2009, lawmakers in the Mexican state of Veracruz accused large-scale hog and poultry operations of being breeding grounds of a pandemic swine flu, although they did not present scientific evidence to support their claim. A swine flu which quickly killed more than 100 infected persons in that area, appears to have begun in the vicinity of a Smithfield subsidiary pig CAFO (concentrated animal feeding operation).

Animal Health Impact

Confinement and overcrowding of animals results in a lack of exercise and natural locomotory behaviour, which weakens their bones and muscles. An intensive poultry farm provides the optimum conditions for viral mutation and transmission – thousands of birds crowded together in a closed, warm, and dusty environment is highly conducive to the transmission of a contagious disease. Selecting generations of birds for their faster growth rates and higher meat yields has left birds' immune systems less able to cope with infections and there is a high degree of genetic uniformity in the population, making the spread of disease more likely. Further intensification of the industry has been suggested by some as the solution to avian flu, on the rationale that keeping birds indoors will prevent contamination. However, this relies on perfect, fail-safe biosecurity – and such measures are near impossible to implement. Movement between farms by people, materials, and vehicles poses a threat and breaches in biosecurity are possible. Intensive farming may be creating highly virulent avian flu strains. With the frequent flow of goods within and between countries, the potential for disease spread is high. Confinement and overcrowding of animals' environment presents the risk of contamination of the meat from viruses and bacteria. Feedlot animals reside in crowded conditions and often spend their time standing in their own waste. A dairy farm with 2,500 cows may produce as much waste as a city of 411,000 people, and unlike a city in which human waste ends up at a sewage treatment plant, livestock waste is not treated. As a result, feedlot animals have the potential of exposure to various viruses and bacteria via the manure and urine in their environment. Furthermore, the animals often have residual manure on their bodies when they go to slaughter.

Confinement at high stocking density requires antibiotics and pesticides to mitigate the spread of disease and pestilence exacerbated by these crowded living conditions. In addition, antibiotics are used to stimulate livestock growth by killing intestinal bacteria. According to a February 2011 FDA report, nearly 29 million pounds of antimicrobials were sold in 2009 for both therapeutic and non-therapeutic use for all farm animal species. The Union of Concerned Scientists estimates that 70% of that amount is for non-therapeutic use

Environmental Impact

Concentrating large numbers of animals in factory farms is a major contribution to global environmental degradation, through the need to grow feed (often by intensive methods using excessive fertiliser and pesticides), pollution of water, soil and air by agrochemicals and manure waste, and use of limited resources (water, energy).

Livestock production is also particularly water-intensive in indoor, intensive systems. Eight percent of global human water use goes towards animal production, including water used to irrigate feed crops.

Industrial production of pigs and poultry is an important source of GHG emissions and is predicted to become more so. On intensive pig farms, the animals are generally kept on concrete with slats or grates for the manure to drain through. The manure is usually stored in slurry form (slurry is a liquid mixture of urine and faeces). During storage on farm, slurry emits methane and when manure is spread on fields it emits nitrous oxide and causes nitrogen pollution of land and water. Poultry manure from factory farms emits high levels of nitrous oxide and ammonia.

Organic pig meat production has a lower global warming potential per kg than does intensive pig meat production. The energy input for free-range poultry meat and eggs is higher than for factory-farmed poultry meat and eggs, but GHG emissions are lower.

Environmental impacts of factory farming can include:

- Deforestation for animal feed production
- Unsustainable pressure on land for production of high-protein/high-energy animal feed
- Pesticide, herbicide and fertilizer manufacture and use for feed production
- Unsustainable use of water for feed-crops, including groundwater extraction
- Pollution of soil, water and air by nitrogen and phosphorus from fertiliser used for feed-crops and from manure
- Land degradation (reduced fertility, soil compaction, increased salinity, desertification)
- Loss of biodiversity due to eutrophication, acidification, pesticides and herbicides
- Worldwide reduction of genetic diversity of livestock and loss of traditional breeds
- Species extinctions due to livestock-related habitat destruction (especially feed-cropping).

Animal Welfare Impact

Animal welfare impacts of factory farming can include:

*   Close confinement systems (cages, crates) or lifetime confinement in indoor sheds
*   Discomfort and injuries caused by inappropriate flooring and housing
*   Restriction or prevention of normal exercise and most of natural foraging or exploratory behaviour
*   Restriction or prevention of natural maternal nesting behaviour
*   Lack of daylight or fresh air and poor air quality in animal sheds
*   Social stress and injuries caused by overcrowding
*   Health problems caused by extreme selective breeding and management for fast growth and high productivity
*   Reduced lifetime (longevity) of breeding animals (dairy cows, breeding sows)
*   Fast-spreading infections encouraged by crowding and stress in intensive conditions
*   Debeaking (beak trimming or shortening) in the poultry and egg insustry to avoid pecking in overcrowded quarters
*   Forced and over feeding (by inserting tubes into the throats of ducks) in the production of foie gras.

Breeding Beef Cattle

There are many things to consider when raising cattle, particularly beef cattle, and one of these is actually breeding beef cattle. Many cattle farmers spend a certain amount of their time trying to breed their cattle not only to be able to increase the herd but to also produce high quality, healthy, and close-to-perfect calves.

They know that consumers and other cattle breeders only want the best in their herd, so breeding beef cattle should be a major addition to one's agenda when engaged in cattle farming. Unfortunately, some farmers tend to forget the most basic tips and techniques to breeding high quality cattle and have failed to renew the expenses made for ill-fitted cows and calves with health problems at birth.To be able to avoid this, here are some helpful tips on how to effectively breed beef cattle:

Breeding beef cattle means knowing how to balance between genetics and herd management. With this in mind, remember to be selective with one's own herd by learning how to distinguish the good animals from those with health and bone structure problems. Keeping the best and raising the best is the key to successful breeding because it not only produces cows with the best meat but also reduces the time, effort, and money spent on work at breeding time.

Bulls and heifers should be examined once in a while, especially when the cows are starting to show pregnancy signs and are ready to conceive. Try certain techniques to increase your chances of producing well-rounded cattle. For example, make sure to choose heifers with large pelvises and a bull whose mother also had a large pelvis.

There are two ways to go for when breeding beef cattle depending on what your preferences are. You can either artificially inseminate the cows or keep a good bull to breed with your cows.

If you've chosen the second method, have your bull/s examined for STDs, health problems, and once in a while for general physical check-ups, sperm count, and scrotum circumference to be sure that your cows will easily conceive.

Choosing the bulls for breeding also means choosing those that have been known to produce well-rounded calves. You can go around, ask other cattle farmers, and see if they have good bulls up for sale that you can keep in your herd.

Top Five Ways for Breeding Finches

Anyone that has finches for pets will eventually consider breeding their birds. The delight you get from your pets makes you want to see your birds have little ones, and watch them grow.

To begin breeding finches you need to get your birds a nest. Some people provide store-bought nests for their pets, some people construct the nests for them, and some people will provide the animals with the materials to build their own nests. Do not be upset if your birds do not like the first nest you give them, this is absolutely normal. Just remove the one they did not like and try another one. Continue to replace them until the birds accept one, and after that you will always know which nest to provide for your breeding finch.

Zebra finch breeders often provide their birds with wicker basket nests that are full of material to encourage their breeders. When these animals breed in the wild they will look for a place where they can hide completely, and this makes them feel safe. If you give them the ability to feel safe this will encourage breeding.

If your pets are not breeding you can try buying a heater from the pet supply, like they place in reptile cages; the heater will make the animals feel better and be healthier. Healthy birds are more likely to breed than sick ones.

You can try providing your animals with plenty of material to put in their nest. Sisal fiber is an excellent choice in materials as the finches breeders who do this on a regular basis will tell you.

One of the best things you can do for your pets is to make their environment as close to the natural environment they would have been in. They need to feel safe and secure, and they need to have what they consider a good food supply. This is not hard to accomplish, you can keep the feeder full for the animals, and plants make them feel like they have places to hide from danger. Buy fake plants for this purpose, or you can try providing them with cover by placing a blanket on the sides of their cage.

If you have children in your home, do not let them bump the cage a lot, or poke their fingers or other objects through the bars. If the birds are frequently disrupted by their cage being jarred around, then they will not feel safe and secure.

When you notice the male fussing with the nest, and then doing their mating song and dance, you should notice an egg in the nest in the next couple of days. The female may not begin to sit on the egg until she has laid 8 eggs. This is called a clutch of eggs, and she can only lay one per day. After they begin to sit on the eggs you will have hatching in 10 to 18 days. The parents will give the young all the care they need.

Learn the joy of Finch companionship and how to buy, keep and raise healthy Finches. Come find all the information you need at Zebra Finch Breeders. I am Ralph Siskin and I have been raising and learning about Finches for quite some time. I love these birds and want to share what I know to other like minded Finch lovers or people just interested in birds.

Breeding Dogs-The Revolutionary New Methods

Dog breeding is a controversial subject. Many people have heard terrible stories of female dogs being bred time after time for profit and then being abandoned by their owners when they are no longer able to have puppies. Bearing this in mind, it is easy to see why some people may view dog breeding as a negative thing. However, many breeders have a genuine love of the breed(s) they work with and this is what motivates them to get into dog breeding in the first place.

So, if you are considering breeding dogs, the questions you need to ask yourself are-What is motivating you? Do you wish to become a professional breeder, or just mate the family pet? It is not enough to say you simply wish to make some money or you think puppies are cute-puppies are cute, but they are also extremely hard work and cost an absolute fortune to rear.

Consider the cost of finding a suitable stud for your bitch, then all the costs associated with pregnancy: vet bills, adequate food, etc. This will be expensive enough, but any such costs will increase ten-fold once the pups are born (bearing in mind that some breeds can have up to 12 babies). Moreover, if a bitch needs a caesarean to deliver her brood, the fee can be astronomical and you need to be prepared for this.

For professional breeders, mating dogs is a full-time job. They have an incredible commitment to the job and an inherent love of dogs, and this is why they go into the job. Obviously, money is a motivating factor for such people-they have to make a living-but it is not THE motivating factor.

A further thing you need to consider if you are to breed your bitch is time. Not only does a pregnant dog cost money, she requires adequate exercise and will need several short walks every day. Also, in the days leading up to the birth (known as 'the whelping period'), you will need to be constantly present for when labour starts. The newborn pups will also need close attention once they arrive, and so it is imperative that you, or someone else in the family, has the time to devote to all of this.

A responsible dog breeder always makes sure that their dog(s) receive all the relevant innoculations and the pups also. If there are 10-12 pups, this can cost a fortune, but it is important that it is done in order to protect the puppies' health. Another consideration in the mess that pups create-initially, you will need a box big enough to house them, and bedding will need to be changed several times every day. When the pups are old enough to walk about, they will not be house trained and will do their business anywhere-therefore, you need to place them in a room with a hard floor that can be easily cleaned. The pups will also need a secure outside space in which they can run, play and get exercise.

Finally, if you do decide to mate your bitch, do your research on the particular breed-ignorance may put your dog and their babies at risk. Also, when you come to sell the puppies, buyers will ask all sorts of questions and you need to be able to answer them. If you are thinking of becoming a professional breeder, you may need a licence-contact your local government office for advice.

┌─────┐
│ **3** │
└─────┘

Microorganisms in Breeding

Microorganisms as Foods

Currently, there are no known commercial products containing live genetically modified microorganisms (GMMs) on the market. In the United Kingdom, GM yeast for beer production has been approved since 1993, but the product was never intended to be commercialized (NCBE 2005). Other microorganisms used in foods (which are in the R&D phase) include starter fermentation cultures for various foods (bakery and brewing), and lactic acid bacteria in cheese. R&D is also aimed at minimizing infections by pathogenic microorganisms and improving nutritional value and flavour.

Attempts have been made to genetically modify ruminant microorganisms for protecting livestock from poisonous feed components. Microorganisms improved by modern biotechnology are also under development in the field of probiotics, which are live microorganisms that, when consumed in adequate amounts as part of food, confer a health benefit on the host (FAO/WHO 2001c).

Food Ingredients, Processing Aids, Dietary Supplements and Veterinary Chemicals Derived from GM Microorganisms

Many enzymes used as processing aids in food and feed production are derived through the use of GMMs (European Commission 2004). This means that the GM microorganisms are inactive, degraded or removed from the final product. GM yeasts, fungi and bacteria have been in commercial use for this purpose for over a decade. Examples include: alpha-amylase for bread-making, glucose isomerase for fructose production, and chymosin for cheese-making. Most of the microorganisms modified for food processing are derivatives of microorganisms used in conventional food biotechnology. GMMs are also permitted in a number of countries for the production of micronutrients, such as vitamins and amino acids used for food or dietary supplement purposes. An example is the production of carotenoids (used as food additives, colourants or dietary supplements) in GM bacterial systems. For animal husbandry, veterinary products such as bovine somatropin, used for increasing milk production, have been developed using genetic engineering. Bovine somatropin has been on the market in several countries for over a decade.

The technique of protein engineering aims at altering the genetic, and thus amino acid, sequence of enzymes. Hitherto, protein engineering has not been used extensively in enzyme production. R&D in this area aims to change enzyme characteristics, e.g. improve temperature or pH stability. Enzymatic processing often replaces existing chemical reactions. In many instances, this results in lowered energy consumption and less chemical waste. Over the past 50 years, advances in genetics and molecular biology have enabled the development and commercial release of GMOs with traits that transcend the species barrier. The traits borne by GMOs may potentially bring significant benefits to the production of food.

Currently, the most frequently commercialized GMOs are crops of soybean, maize and cotton. GM soybean dominates plantings of GM crops, followed by GM maize and GM cotton. GM crops are estimated to cover almost 4% of total global arable land. Agronomic traits are the most prominent traits introduced in GM crops. In the near future, agronomic traits will continue to dominate new varieties of GM crops. However, over the medium term, a small but increasing proportion of GM crops will contain changes in quality and nutritional traits. While fast-growing GM salmon and GM cattle expressing increased levels of protein are in an advanced stage of development, most other transgenic animals for food use are still in the early stages of R&D. Many food-processing aids (enzymes) produced through the use of GM microorganisms have been on the market for over a decade, and are used in a wide variety of processed foods. Hitherto, no live GM food microorganisms as such have been introduced onto the market.

History of Risk Assessment of GMOs

When new foods (crop varieties, animal breeds or microorganisms) are developed by traditional breeding methods, they are usually not subject to specific pre-or postmarket risk or safety assessment by national authorities or through international standards. This is in contrast to requirements introduced for GMOs and GM foods.

The concept of risk assessment of GMOs was first discussed at the Asilomar Conference in 1975 (Fredrickson 1979; Talbot 1983). The discovery of recombinant DNA had raised concerns among researchers regarding the potential creation of recombinant viruses whose escape would threaten public health. Fourteen months after a voluntary moratorium on research involving recombinant DNA techniques, guidelines for the physical and biological containment of riskier experiments were drafted and agreed. These guiding principles were the basis of the USA guidelines for research in modern biotechnology developed in 1976 by the National Institutes of Health Recombinant DNA Advisory Committee. Other countries were soon to follow (OECD 1986).

Early regulatory requirements were intended to prevent the accidental release of microorganisms from research facilities. In continuation of this, regulation for contained use and deliberate release of GMOs was developed, e.g. EU regulations in 1990. These guidelines elaborated a premarket human-health and environmental-safety assessment requirement for all GMOs and GM foods on the basis that they are novel and have no history of safe food or environmental use.

Many countries have since established specific premarket regulatory systems requiring the rigorous assessment of GMOs and GM foods before their release into the environment and/or use in the food supply. A summary of some national and international legislation is available on the OECD Internet site (OECD 2005).

While many national regulatory bodies base their safety assessment of GMOs and GM foods on shared concepts, differences in regulatory systems have led to disagreements and confusion in their deployment. While the terms 'safety assessment' and 'risk assessment' are often used interchangeably in some literature, these are two clearly different, but interlinked processes.

To provide international consistency in risk analysis of GMOs and GM foods which incorporates risk assessment, management and communication components, a number of international regulatory and standard-setting bodies have introduced uniform standards. These include standards for human-health and environmental-safety assessment of GMOs and GM foods, and notification of their movement across national borders. The objective of uniform global standards for risk assessment would be challenging as countries are bound to reach different decisions on the scope of the assessment, particularly the resolution of whether or not to include social or economic aspects.

International regulatory systems covering GM food safety (Codex *Principles*) (CAC 2003b) and environmental safety (Cartagena Protocol on Biosafety) (CBD 2000) came into force in 2003.

The concept that allows for the comparison of a final product with one having an acceptable standard of safety is an important element of a GM food safety assessment. This principle was elaborated by FAO, WHO and OECD in the early 1990s and referred to as 'substantial equivalence' (FAO/WHO 1990). The principle suggests that GM foods can be considered as safe as conventional foods when key toxicological and nutritional components of the GM food are comparable to the conventional food (within naturally occurring variability), and when the genetic modification itself is considered safe (OECD 1993). However, the concept has been criticized by some researchers (Millstone et al. 1999). At a *Joint FAO/WHO consultation on foods derived from biotechnology* held in 2000, it was acknowledged that the concept of substantial equivalence contributes to a robust safety assessment, but it was also clarified that the concept should represent the starting point used to structure the safety assessment of a GM food relative to its conventional counterpart (FAO/WHO 2000). The consultation concluded that a consideration of compositional changes should not be the sole basis for determining safety, and that safety can only be determined when the results of all aspects under comparison are integrated.

This study does not cover aspects of occupational health which are often addressed in regulations dealing with the safety of work with GMOs in contained areas. It should also be noted that the adventitious presence of non-approved products of modern biotechnology among the approved is not within the scope of this study.

Assessment of the Impact of GM foods on Human Health

The Codex Alimentarius Commission[2] (CAC, or Codex) adopted the following texts in July 2003: *Principles for the risk analysis of foods derived from modern biotechnology*; *Guideline for the conduct of food safety assessment of foods derived from recombinant-DNA plants*; and *Guideline for the conduct of food safety assessment of foods produced using recombinant-DNA microorganisms*. The last two texts are based on the *Principles* and describe methodologies for conducting safety assessments for foods derived from recombinant-DNA plants and microorganisms, respectively (CAC 2003b,c,d).

The premise of the *Principles* dictates a premarket assessment, performed on a case-by-case basis and including an evaluation of both direct effects (from the inserted gene) and unintended effects (that may arise as a consequence of insertion of the new gene). The Codex safety assessment principles for GM foods require investigation of:

(a) direct health effects (toxicity);

(b) tendency to provoke allergic reactions (allergenicity);

(c) specific components thought to have nutritional or toxic properties;

(d) stability of the inserted gene;

(e) nutritional effects associated with the specific genetic modification; and

(f) any unintended effects which could result from the gene insertion.

Codex principles do not have a binding effect on national legislation, but are referred to specifically in the *Agreement on the Application of Sanitary and Phytosanitary Measures* of the World Trade Organization, and are often used as a reference in the case of trade disputes.

The 2003 *Expert consultation on the safety assessment of foods derived from GM animals, including fish* (CAC 2003a) formed the opinion that to further develop the risk-assessment process with current scientific knowledge, integrated toxicological and nutritional evaluations should be conducted in order to identify food-safety issues that may need further investigation.

Potential Direct Effects on Human Health

The potential direct health effects of GM foods are generally comparable to the known risks associated with conventional foods, and include, for example, the potential for allergenicity and toxicity of components present, and the nutritional quality and microbiological safety of the food. As mentioned above, many of these issues have not traditionally been specifically assessed for conventional food; but in one area — toxicity of food components — there is ample experience related to the use of animal experiments to test potential toxicity of targeted chemical components. However, the intrinsic difficulty in testing whole foods, as opposed to specific components, in animal feeding experiments have resulted in the development of alternative approaches for the safety assessment of GM foods.

The safety assessment of GM food follows a stepwise process aided by a series of structured questions. Factors taken into account in the safety assessment include:

- identity of gene of interest, including sequence analysis of flanking regions and copy number;
- source of gene of interest;
- composition of GMO;
- protein expression product of the novel DNA;
- potential toxicity;
- potential allergenicity; and
- possible secondary effects from gene expression or the disruption of the host DNA or metabolic pathways, including composition of critical macronutrients, micronutrients, antinutrients, endogenous toxicants, allergens and physiologically active substances.

A series of FAO/WHO expert consultations held in 2000, 2001 and 2003 recognized that animal studies can be of help but that there are practical difficulties in obtaining meaningful information from conventional toxicology testing, especially with whole-food studies in laboratory animals (where the appropriate animal diet is a factor that needs to be assured) (FAO/WHO 2000, 2001b, 2003a). The consultations also noted that very little is known about the potential long-term effects of any foods. At present, there is no conclusive information available on the possible health effects of modifications which would significantly change the nutritional characteristics of any food, such as nutritionally enhanced foods.

Potential Unintended Effects of GM foods on Human Health

Unintended effects, such as elevated levels of antinutritional or toxic constituents in food, have on occasion been characterized in conventional breeding methods, e.g. glycoalkaloid levels in potatoes. Organisms derived from conventional breeding methods, including tissue cultures, may have a somewhat enhanced possibility for genetic (and epigenetic — environmentally induced changes that affect the expression of a gene without changing the DNA sequence) instabilities, such as the activity of mobile elements and gene-silencing effects (FAO/WHO 2003a). These effects could increase the probability of unintended pleiotropic effects (affecting more than one phenotypic trait), e.g. increased or decreased expression of constituents or possibly modifications in expressed proteins, as well as epistasis (the interaction of the inserted gene with other genes).

It has been argued that random insertion of genes in GMOs may cause genetic and phenotypic instabilities (Ho 2002) but, as yet, no clear scientific evidence for such effects is available. A better understanding of the impact of natural transposable elements on the eukaryotic genome may shed some light on the random insertion of sequences.

Gene expression in conventional and GM crops is subject to environmental influences. Environmental conditions such as drought or heat can stimulate some genes; turning the expression up or down. The assessment of potential synergistic effects is necessary in the risk assessment of organisms derived from gene stacking, i.e. breeding of GMOs containing genetic constructs with multiple traits. Internationally agreed procedures for the assessment of such organisms are desirable.

Unintended effects can be classified as insertional effects, i.e. related to the position of insertion of the gene of interest, or as secondary effects, associated with the interaction between the expressed products of the introduced gene and endogenous proteins and metabolites. There is common agreement that targeted approaches, i.e. the measuring of single compounds, is very useful and adequate to detect such effects, as has been done with conventionally bred products. To enhance and improve the identification and analyses of these unintended effects, profiling methods have been suggested. This untargeted approach allows detection of unintended effects at the mRNA (microarray), protein (proteomic) and metabolite (metabolomic) level. It still remains to be seen which of these techniques (once validated) would be useful for routine risk-assessment purposes.

Unintended effects were specifically addressed by the FAO/WHO *Expert consultation on the safety aspects of genetically modified foods of plant origin* (FAO/WHO 2000) and the Codex *Principles for the risk analysis of foods derived from modern biotechnology* (CAC 2003b). These consultations noted that there is a need to establish the consequences of natural baseline variations, the effects of growing conditions and environmental influences, and the ways to interpret safety-relevant data from profiling techniques. Adequate methods for the assessment of potential, unintended effects need to be evaluated for specific GMOs case by case, where the assessment already aims to consider unintended toxic and antinutritional factors through analysis of proximal constituents and GM characteristics.

As profiling methods are not in use in routine risk assessment, the second step in the comparative safety assessment has been suggested as a measure for identifying and characterizing any unintended effects that may be associated with complex foods.

Potential Human-health effects from Horizontal Gene Transfer

Natural genetic transformation has been found to occur in different environments, e.g. in food (Kharazmi et al. 2003). In addition, it has been shown that ingested DNA from food is not completely degraded by digestion, and that small fragments of DNA from GM foods can be found in different parts of the gastrointestinal tract. As the consequences of horizontal gene transfer (HGT) may be significant in some human-health conditions, the potential for HGT needs to be part of the risk assessment of GM food.

FAO/WHO consultations (FAO/WHO 2001b) have also discussed the potential risks of gene transfer from GM foods to mammalian cells or gut bacteria. These panels have suggested that it may be prudent in a food-safety assessment to assume that DNA fragments survive in the human gastrointestinal tract and can be absorbed by either the gut microflora or somatic cells lining the intestinal tract. It was agreed that the assessment needs to take into account a number of factors including, but not limited to, the specific characteristics encoded by the DNA sequences, the characteristics of the receiving organism, and the selective conditions of the local environment of the receiving organisms.

Some scientists have pointed to the present methodological limitations of a comprehensive scientific evaluation of this problem (mainly because of estimations that only approximately 1% of naturally existing bacteria can be cultured, and therefore analysed). Discussion also addresses the consequences of a rare probability of a transfer event against the high numbers of bacteria and genes available for transfer.

The DNA construct used to change the genetic composition of a recipient organism should be considered within an assessment, especially if the gene or its promoter (e.g. cytomegalovirus promoter) (Ho et al. 2000) has been derived from a viral source. Sequences unrelated to the target gene could be introduced as part of the construct (FAO/WHO 2003a). Inadvertent introduction of such sequences into the germ-line of a GM animal not only has the potential for creating unintended genetic damage, but can also contribute by recombination to the generation of novel infectious viruses. A well-known example is the generation of a replication-competent murine leukaemia virus during the development of a vector containing a globin gene (Purcell et al. 1996).

The horizontal transfer of recombinant genetic material to microorganisms has demonstrated an enhanced stability of DNA under certain conditions (Lorenz and Wackernagel 1987). Natural transformation of DNA to bacteria involves the active uptake of extracellular DNA by bacteria in a status of competence or in rare, illegitimate recombination events. The probability of such an event occurring appears to be extremely low, and very much related to the genes, constructs and organisms in question.

The FAO/WHO expert panels concluded that horizontal gene transfer is a rare event that cannot be completely discounted, and that the consequences of such transfer should be considered in a safety assessment. The panels encouraged the use of recombinant DNA without antibiotic-resistance genes (particularly those that could interfere with human or animal therapies), or any other sequences which could stimulate transfer.

The panels also discouraged the use of any unnecessary DNA sequences, including marker genes in the genetic construct (FAO/WHO 2001b, 2003a). The safety assessment of a genetic construct should also examine the included marker genes. Commonly used marker genes code for antibiotic resistance. Risk assessment of these selectable genes should focus on gene transfer to microorganisms residing in the gastrointestinal tract of humans or animals. As the potential of this gene transfer cannot be completely ruled out, the safety assessment should also consider information on the role of the antibiotic in human and veterinary medical uses.

Potential Immune Responses and Allergenicity Induced by GM foods

Food allergies or hypersensitivities are adverse reactions to foods triggered by the immune system. Within the different types of reactions involved, non-immunological intolerances to food and reactions involving components of the immune system need to be differentiated. The former may invoke reactions such as bloating or other unpleasant reactions, but are thought not to involve the immune system and called 'food intolerances'.

Allergic reactions to traditional foods are well known. The major food allergens are proteins in and derived from eggs, fish, milk, peanuts, shellfish, including crustaceans and molluscs (e.g. clams, mussels and oysters), soy, tree nuts (e.g. almonds, Brazil nuts, cashews, hazelnuts/filberts, macadamia nuts, pecans, pine nuts, pistachios and walnuts) and wheat. Whereas the groups of main allergens are well known and advanced testing methods have been elaborated, traditionally developed foods are not generally tested for allergens before market introduction.

The application of modern biotechnology to crops has the potential to make food less safe if the newly added protein proves to cause an allergic reaction once in the food supply. A well-known case is the transfer of a gene encoding a known allergen, the 2S-Albumin gene from the Brazil nut, to a previously safe soybean variety. When the allergenic properties of the transgenic soybean were tested, sera from patients allergic to Brazil nuts cross-reacted with the transgenic soybean (Nordlee et al. 1996). For this reason, a commercial product was never pursued. On the other hand, the introduction of an entirely new protein that has not been previously found in the food chain represents a different case.

In the first case, guidelines for assessing foods with known allergens are clear. The second case is more difficult to assess because there is no definitive test to determine the potential allergenicity of a novel protein. Instead, several risk factors provide a rough guide as to the likelihood of allergenicity.

Risk-assessment protocols for food allergy examine four elements: (1) allergenicity assessment (is the food or elements in the food a potential allergen); (2) dose response assessment (is there a safe concentration of the allergen); (3) exposure assessment (how likely is it that people will encounter the allergen); and (4) susceptible subpopulations (how do those prone to allergy react to this new food). Elements of an allergenicity assessment include a comparison of the sequence of the transferred gene (including the flanking regions at the site of insertion) with sequence motifs of allergenic proteins from databanks, an evaluation of the stability of the newly expressed proteins against digestion, and animal and immune tests, as appropriate. Absence of sequence similarity with allergenic protein epitopes, and low stability under acidic or proteolytic conditions, do not preclude the presence of a potential allergen. There are proven incidents which have contradicted the general rules, e.g. where small modifications in a protein sequence determine allergenicity (Ferreira et al. 1996). Allergenicity prediction using protein-sequence motifs identified from a new allergen database has been proposed as a new and superior strategy for identifying potential allergens. Some experts consider that the use of sera from polysensitized patients is important for the testing of allergenicity. Areas of improvement of risk assessment of allergens include mechanistic studies of animal models and genomic techniques.

FAO/WHO expert panels (FAO/WHO 2001a) have established protocols for evaluating the allergenicity of GM foods on the basis of the weight of evidence. The strategy adopted is applicable to foods containing a gene derived from either a source known to be allergenic or a source not known to be allergenic. The panels have, however, discouraged the transfer of genes from known allergenic foods unless it can be demonstrated that the protein product of the transferred gene is not allergenic. These principles have been applied by many regulatory agencies assessing the safety of GM foods and have provided the basis for Codex guidelines for the safety assessment of foods derived from biotechnology (CAC 2003c,d).

The cellular basis of immune responses is not completely understood, and a better understanding of the interaction of the immune system and foods in general is required in order to decipher whether specific GM foods may have impacts on the immune system apart from allergenicity. The impact of cell-mediated reactions (without involvement of immunoglobulin E antibodies) on hypersensitivity reactions elicited by foods is a matter of current research.

Safety Aspects of Food Derived from GM Animals

Genetically modified animals have mainly been produced for biomedical research purposes. To date, no GM food animals have been introduced onto international markets. But GM food animals such as fish can be expected in the near future.

In principle, the assessment of food and feed safety for GM animals follows the general principles of the assessment of GMOs outlined above. However, the specificities of the introduction of transgenes into animals, often using viral constructs for introduction into the germ-line, need distinct consideration. A 2003 report of the Pew Initiative on Food and Biotechnology (PIFB 2003) reviewed techniques for the production, uses and welfare of GM animals, as well as safety aspects.

The risk assessment of foods derived from GM animals needs to be undertaken, as for other GM foods, on a case-by-case basis (CAC 2001a). This includes an assessment of potential recombination of viral vectors used for transformation with wild-type viruses (Mikkelsen and Pedersen 2000), especially in poultry, where potential incomplete digestion could lead to intestinal uptake of orally administered proteins, and an assessment of peptide expression that may have hormonal activity (e.g. in fish).

The FAO/WHO expert consultation on the *Safety assessment of foods derived from GM animals, including fish* held in 2003 addressed the key issues for food safety and evaluated the extent of scientific knowledge with regard to hazard identification and characterization unique to transgenic animals (FAO/WHO 2003a).

Phenotypic analysis: Because of their size, and limitations in the generation process, there will in general be few initial founders for screening of GM animals, meaning that information on the variation range between animals with the same genetic modification will be rather limited. This will make interpretation of differences difficult. Furthermore, a selection of the edible tissues and products to be analysed has to be made for the different animal species. In specific cases, phenotypic analysis may also be advisable after processing or, for fish, during the various stages of spoilage. For example, adverse biogenic amines can be formed during spoilage in salmon, tuna, herring and other fish species. Similarly, formaldehyde may be produced in spoiled shrimp, cod, hake and many other species.

Compositional analysis: Background data on the natural variation for individual constituents in different tissues need to be generated. Data in existing databases must be evaluated for their quality and value for use in comparative compositional analysis.

Safety Aspects of Foods Derived From or Produced with GMMs

The production of food additives or processing aids using GMMs, where the microorganism is not a part of the food, has become an important and generally well-accepted technology, with a significant number of such products on the market (Ross et al. 2002). Experience with the purification of proteins in the biomedical field suggests that well-standardized purification protocols are of central importance for the safety of these products. Where the GMMs are a part of the food matrix (e.g. starter culture containing live or sterilized microbes), certain criteria were established in 2001 by a *Joint FAO/WHO expert consultation on foods derived from biotechnology* (FAO/WHO 2001b) for assessing the risks that may be associated with the preparation of such foods. These include the genetic constructs (vectors) used in the GM microorganisms, the pathogenic potential of GMM, and the detrimental effects of a potential gene transfer (considering a higher incidence for gene transfer (Salyers et al. 2004) and the various mechanisms involved).

For GMMs used in foods (e.g. in fermented foods or in functional food preparations), the ensuing risk assessment ought to focus on the effects of a possible interaction between the GMM and endogenous intestinal microflora and the potential immune-stimulatory or immunomodulatory effects of the microorganisms in the event the gastrointestinal tract is colonized (FAO/WHO 2001b). Small regulatory elements derived from viral DNA are commonly used to drive the expression of transgenes in GMOs. Viral-DNA constructs are sometimes used as transgenes to establish resistance against viral pests, as they express viral proteins that confer viral resistance on plants. Some scientists suggest that the potential interaction of GM viral constructs with related wild-type viruses needs to be part of the risk assessment, to evaluate the potential of new viral pest strains evolving through mechanisms of recombination. Insertion of viral vectors into functionally important genes of recipient patients in the field of biomedicine has been reported, and whereas such vectors are not commonly used in food production, this evidence indicates the limited understanding of mechanisms directing insertion of genetic constructs (Check 2003).

Safety Aspects of Foods Derived from Biopharming

The potential to produce human proteins in animals has resulted in great interest in new possibilities for human health, but also in efforts to establish appropriate risk-assessment methods. The biosafety aspects of molecular 'farming' (or 'pharming') can be divided into two major groups: the potential spread of transgenes; and the potential negative effects of the expressed protein on the environment and the consumer. Practices and guidelines ensuring effective separation of 'biopharming' are being investigated. Experts agree that the risk assessment should ensure that proteins designed to produce pharmaceutical products, e.g. in the animal's milk, cannot find their way to other parts of the animal's body, possibly causing adverse effects.

Potential Effect of GMOs on Human Health Mediated Through Environmental Impact

Work on environmental-health indicators (von Schirnding 2002) suggests that various agricultural practices have direct and indirect effects on human health and development. Hazards can take many forms — wholly natural in origin, or derived from human activities and interventions. The need to assess indirect effects of the use of GMOs in food production has been emphasized by many countries. Potential environmental-health hazards from the release of GMOs into the environment have been discussed in a report by WHO and the Italian Environment Protection Agency, in which health effects have been analysed "as an integrating index of ecological and social sustainability" (WHO/EURO– ANPA 2000). For example, the production of chemicals or enzymes from contained GM microorganisms (e.g. chemicals, pharmaceuticals or food additives) have contributed significantly to decreases in the amount of energy use, toxic and solid wastes in the environment, thereby significantly enhancing human health and development (CBD 2003).

A further example of beneficial human/environmental outcomes of introducing GM crops is the reduction in the use, environmental contamination and human exposure to pesticides demonstrated in some areas. This has occurred especially through the use of pesticide-resistant *Bt* cotton, which has been shown to decrease pesticide poisoning in farm workers (Pray et al. 2002).

Outcrossing of GM plants with conventional crops or wild relatives, as well as the contamination of conventional crops with GM material, can have an indirect effect on food safety and food security by contamination of genetic resources. Although initial concerns about introgression of transgenic DNA into traditional landraces of maize in Mexico arose as a result of the findings of transgenic DNA in such landraces in 2000 recently published results from samples taken during a broad, systematic survey in 2003 and 2004 in the same region shows no transgenes in these landraces (detection limit approx. 0.01%) (Ortiz-García et al. 2005). Still, the potential for introgression remains a possibility and risk-mitigation measures are being considered.

Both outcrossing and contamination characteristics are dependent on the pollination and distribution characteristics of pollen and seeds of the specific plant. In the USA, GM 'Starlink' maize was not approved for food use, but unintentionally started to appear in maize food products. This example demonstrated the problem of contamination and highlighted the potential for unintended impacts on human health and safety. In the case of Starlink maize, full segregation of GM varieties not intended for food use and other varieties of the same crop species could not be achieved.

Improved molecular methods for containment of the transgenes as well as farm management measures are under discussion, e.g. isolation distances, buffer zones, pollen barriers, control of volunteer plants, crop rotation and planting arrangements for different flowering periods, and monitoring during cultivation, harvest, storage, transport and processing.

The likelihood of GM animals entering and persisting in the environment will vary among taxa, production systems, modified traits, and receiving environments. The spread and persistence of GM fish and shellfish — or their transgenes — in the environment could be an indirect route of entry of GM animal products into the human food supply. This is because escaped individuals or their descendents could subsequently be captured in fishing for those species. Similar mechanisms might apply for poultry such as ducks and quail that are subject to sport or subsistence harvest. Live transport and sale of GM fish and poultry pose another route for the escape of GM animals and their entry into the environment.

GMOs and Environmental Safety

In many national regulations, the elements of environmental risk assessment (ERA) for GM organisms include the biological and molecular characterizations of the genetic insert, the nature and environmental context of the recipient organism, the significance of new traits of the GMO for the environment, and information on the geographical and ecological characteristics of the environment in which the introduction will take place.

The risk assessment focuses especially on potential consequences for the stability and diversity of ecosystems, including putative invasiveness, vertical or horizontal gene flow, other ecological impacts, effects on biodiversity and the impact of the presence of GM material in other products (Connor et al. 2003). Different approaches in the ERA regulations of different countries have often resulted in different conclusions on the environmental safety of certain GMOs, especially where the ERA focuses not only on the direct effects of GMOs, but also addresses indirect or long-term effects on ecosystems, e.g. impact of agricultural practices on ecosystems (FAO/WHO 2004).

Internationally, the concept of 'familiarity' was developed also in the concept of environmental safety of transgenic plants. The concept facilitates risk/safety assessments, because to be familiar means having enough information to be able to make a judgement of safety or risk (FAO/WHO 2000). Familiarity can also be used to indicate appropriate management practices, including whether standard agricultural practices are adequate or whether other management practices are needed to manage the risk (FAO/WHO 2000). Familiarity allows the risk assessor to draw on previous knowledge and experience with the introduction of plants and microorganisms into the environment and informs appropriate management practices. As familiarity depends also on knowledge of the environment and its interaction with introduced organisms, the risk/safety assessment in one country may not be applicable in another country.

Currently, the Cartagena Protocol on Biosafety (CPB) of the Convention on Biological Diversity is the only international regulatory instrument which deals specifically with the potential adverse effects of GMOs (known as living modified organisms (LMOs) under the Protocol) on the environment, taking also into account effects on human health (CBD 2000). The Protocol covers transboundary movements of any GM foods that meet the definition of an LMO. Annex III of the Protocol specifies general principles and methodology for risk assessment of LMOs. The Protocol establishes a harmonized set of international rules and procedures designed to ensure that countries are provided with the relevant information, through the information exchange system called the Biosafety Clearing House (CBD 2005c). This Internet-based information system enables countries to make informed decisions before agreeing to the importation of LMOs. It also ensures that LMO shipments are accompanied by appropriate identification documentation. While the Protocol is the key basis for international regulation of LMOs, it does not deal specifically with GM foods, and its scope does not consider GM foods that do not meet the definition of an LMO. Furthermore, the scope of its consideration of human-health issues is limited, given that its primary focus is biodiversity, in line with the scope of the Convention itself. Consequently, the Protocol alone (which entered into force on 11 September 2003) is not sufficient for the international regulation of GM foods.

Potential Unintended Effects of GMOs on Non-Target Organisms, Ecosystems and Biodiversity

This study does not focus specifically on the effects that GMOs used in food production may have on the environment. However, these aspects need to be considered in a holistic assessment of GM food production, as environmental effects may indirectly affect human health and development in many ways.

Potential risks for the environment include unintended effects on non-target organisms, ecosystems and biodiversity. Insect-resistant GM crops have been developed by expression of a variety of insecticidal toxins from the bacterium *Bacillus thuringiensis* (*Bt*). Detrimental effects on beneficial insects, or a faster induction of resistant insects (depending on the specific characteristics of the *Bt* proteins, expression in pollen and areas of cultivation), have been considered in the ERA of a number of insect-protected GM crops. Studies on the toxicity of *Bt* maize on the monarch butterfly in the USA indicate that for most commercially available hybrids, the *Bt* expression in pollen is low, and laboratory and field studies show that no acute toxic effects at any pollen density would be encountered in the field (Sears et al. 2001). These questions are considered an issue for monitoring strategies and improved pest-resistance management.

Increased doses of herbicide can be applied post-emergence to herbicide-tolerant crops, thus avoiding routine pre-emergence applications and reducing the number of applications needed. Also, the need for tilling can be reduced under critical soil conditions. In certain agro-ecological situations, such as a high weed pressure, the use of herbicide-tolerant crops has resulted in a reduction in quantity of the herbicides used. However, in other cases, herbicide use has stayed the same or even increased. In other situations, the following have been investigated: potentially detrimental consequences for plant biodiversity, weed shifts to less-sensitive species and development of herbicide resistance, decreased biomass, adverse effects on wildlife such as arthropods or birds, or consequences for agricultural practices, e.g. the use of the ecologically important practice of crop rotation.

Outcrossing: Outcrossing of transgenes has been reported from fields of commercially grown GM plants, including oilseed rape and sugar beet, and has been demonstrated in experimental releases for a number of crops, including rice and maize. Outcrossing could result in an undesired transfer of genes such as herbicide-resistance genes to non-target crops or weeds, creating new weed-management problems.

The consequences of outcrossing can be expected in regions where a GM crop has a sympatric distribution and synchronized flowering period that is highly compatible with a weedy or wild relative species, as demonstrated for rice. In view of the possible consequences of gene flow from GMOs, the use of molecular techniques to inhibit gene flow has been considered and is under development. Isolation distances or future molecular strategies for transgene confinement in transgenic crops may reduce gene flow (Daniell 2002). Stringent isolation measures may be necessary because of complex dispersal mechanisms for certain crops. Gene confinement techniques, e.g. introducing the transgene in plastids which are not inherited paternally, are either not very effective because of gene flow via seeds (Board on Agriculture and Natural Resources 2004; Snow et al. 2004) or they are still in an early stage of development.

GM animals: The possibility that certain genetically engineered fish and other animals may escape, reproduce in the natural environment and introduce recombinant genes into wild populations is raised in a report of a United States Academy of Science study (Board on Agriculture and Natural Resources 2002). Genetically engineered insects (PIFB 2004), shellfish, fish and other animals that can easily escape, are highly mobile and that form feral populations easily, are of concern, especially if they are more successful at reproduction than their natural counterparts. For example, it is possible that if released into the natural environment, transgenic salmon, with genes engineered to accelerate growth could compete more successfully for food and mates than wild salmon, thus endangering wild populations. The use of sterile, all-female genetically engineered fish could reduce interbreeding between native populations and farmed populations (Muir and Howard 2002), a current problem with the use of non-engineered fish in ocean net-pen farming. Sterility eliminates the potential for spread of transgenes in the environment, but does not eliminate all potential for ecological harm. Monosex triploidy is the best existing method for sterilizing fish and shellfish, although robust triploidy verification procedures are essential.

GMMs: Gene transfer from bacteria to bacteria in the soil has been demonstrated in some systems, e.g. for antibiotic-resistance genes (Nwosu 2001), and only a limited number of releases of GMMs (e.g. *Pseudomonas* and *Rhizobium*) has been permitted; mainly to explore the spread and the fate of microorganisms in nature.

Risk assessment in this field is impeded by a number of factors, such as the limited knowledge of microorganisms in the environment (only approximately 1% of soil bacteria are currently described), the existence of natural transfer mechanisms between microorganisms, and the difficulties in controlling their spread.

Status of Methods for Estimating Potential Environmental Entry

Methods by which to reliably characterize potential environmental entry have not yet been standardized. Net-fitness methodology (Muir and Howard 2002) does provide, however, a systematic and comprehensive approach based on contemporary evolutionary and population biology. It involves a two-step process of (1) measuring fitness-component traits covering the entire life-cycle for GM animals, their conventional counterparts, and crosses between the two; and (2) entering the fitness data from step 1 into a simulation model that predicts the fate of the transgene across multiple generations. There is a need to validate the predictions based on this method. Initial experiments are under way to this end (FAO/WHO 2003a).

Regional Specificity in Safety Assessments

Contradictory findings for benefits and risks for the same GM crop may reflect that such effects may be a consequence of different agro-ecological localities or regions. For example, the use of herbicide-resistant crops could potentially be detrimental in a small-sized agricultural area which has extensive crop rotation and low levels of pest pressure. Moderate herbicide uses on these GM plants could be beneficial in other agricultural situations.

At present, no conclusive evidence on environmental advantages or costs can be generalized from the use of GM crops. Consequences may vary significantly between different GM traits, crop types and different local conditions including ecological and agro-ecological characteristics.

In the USA, the overall difference in herbicide use between GM and conventional soybeans ranged from +7 to –40% (1995–1998), with an average reduction of 10%. These changes have been associated with a number of factors including soil type, weed pressure, farm size, management style, prices of different herbicide programmes, and climate (Hin et al. 2001). Potential advantages of *Bt* maize have been widely attributed to regions with a significant pest pressure from the maize borer (Obrycki et al. 2001).

The consequences of outcrossing can produce highly different characteristics, depending on the potentially different recipient plants in different ecological regions (Snow 2002). These observations suggest that risk assessment needs to reflect the regional specificities of the receiving environments in addition to the characteristics of the GMO.

In 1999, the government of the United Kingdom asked an independent consortium of researchers to investigate how growing GM crops might affect the abundance and diversity of farmland wildlife compared with growing conventional varieties of the same crops (Andow 2003). In the largest-ever field trials of GM crops in the world, the researchers compared three GM crops with their conventional counterparts. The crops were sugar-and fodder-beet (considered as a single crop), spring-sown oilseed rape, and maize. The GM crops had been genetically modified to make them resistant to specific herbicides. Other types of GM crops, such as those engineered to be resistant to certain insect pests, were not included in the study. The team found that there were differences in the abundance of wildlife between GM herbicide-tolerant crop fields and conventional crop fields. Growing conventional beet and spring rape was better for many groups of wildlife than growing GM herbicide-tolerant beet and spring rape. There were more insects, such as butterflies and bees, in and around the conventional crops because there were more weeds to provide food and cover. There were also more weed seeds in conventional beet and spring rape crops than in their GM counterparts. Such seeds are important in the diets of some animals, particularly some birds. In contrast, growing GM herbicide-tolerant maize was better for many groups of wildlife than conventional maize. There were more weeds in and around the GM herbicide-tolerant crops, more butterflies and bees at certain times of the year, and more weed seeds.

The researchers stress that the differences they found do not arise just because the crops have been genetically modified. They arise because these GM crops give farmers new options for weed control. That is, they use different herbicides and apply them differently. The results of this study suggest that growing such GM crops could have implications for wider farmland biodiversity. However, other issues will affect the medium-and long-term impacts, such as the areas and distribution of land involved, how the land is cultivated and how crop rotations are managed. These make it hard for researchers to predict the medium-and large-scale effects of GM cropping with any certainty. In addition, other management decisions taken by farmers growing conventional crops will continue to have an impact on wildlife.

Monitoring of the environmental impacts of GM crops in various regions and from investigation over longer time periods may be necessary to conclude on effects and consequences.

Monitoring of Human Health and Environmental Safety

In the future, GMOs may gain wider approval for environmental release, either with or without approval to enter them in the human food supply. In such situations, it will be important to consider whether or not to apply postmarket monitoring for unexpected environmental spread of the GMOs and their transgenes) that may pose food safety hazards. Methods for detection of such GMOs and their transgenes in the environment are likely to involve application of two well-established bodies of scientific methodologies: (1) diagnostic, DNA-based markers; and (2) sampling protocols that are adequate (in terms of statistical power) and cost-effective. However, there is a need to fully develop appropriate protocols for application of these methods to postmarket detection of environmental spread of GMOs and their transgenes. Monitoring can also be helpful to assure confinement of GMOs during R&D (FAO/WHO 2003a).

Postmarket monitoring (or surveillance) of GM foods with respect to direct human-health impacts has been raised at international conferences (Health Canada 2002) and within the Codex Alimentarius Commission. Opinions regarding such monitoring vary from neither necessary nor feasible, to being essential to support and improve the results of a risk assessment and enable an early detection of uncharacterized and unintended hazards. Some have suggested that monitoring of potential long-term effects of GM foods with significantly altered nutritional composition (Amanor-Boadu and Amanor-Boadu 2002) should be mandatory. The *Expert consultation on the safety assessment of foods derived from GM animals* held in 2003 (FAO/WHO 2003a) identified a need for postmarket surveillance, and therefore a product-tracing system, for:

- confirmation of the (nutritional) assessments made during the premarket phase;
- assessment of allergenicity or long-term effects; and
- unexpected effects.

The issue of postmarket surveillance is closely related to risk characterization. In general, potential safety issues should be addressed adequately through premarket studies, as the potential of postmarket studies is currently very limited. Postmarket surveillance could be useful in certain instances where clear-cut questions require, for instance, a better estimate of dietary exposure and/or the nutritional consequences of GMO-derived food.

Tools to identity or trace GMOs or products derived from GMOs in the environment or food-chain are a prerequisite for any kind of monitoring. Detection techniques (such as polymerase chain reaction; PCR) are in place in a number of countries to monitor the presence of GMOs in food, to enable the enforcement of GM labelling requirements, and to monitor effects on the environment. Attempts to standardize analytical methods for tracing GMOs have been initiated (European Commission 2002).

GM foods currently available on the international market have undergone risk assessments and are not likely to present risks for human health any more than their conventional counterparts.

The risk-assessment guidelines specified by CAC are thought to be adequate for the safety assessment of GM foods currently on the international market. Guidelines for environmental risk assessment have been developed under the Convention on Biological Diversity.

The potential risks associated with GMOs and GM foods should be assessed on a case-by-case basis, taking into account the characteristics of the GMO or the GM food and possible differences of the receiving environments.

In the field of potential risks derived from outcrossing or contamination from GM crops, relevant consequences need to be investigated for specific crops, and strategies for risk management need to be explored.

As highlighted in the Codex *Principles for the risk analysis of foods derived from modern biotechnology* (CAC 2003b), the assessment of the potential of GM foods to elicit hypersensitivity reactions should be part of the risk assessment for GM foods. This includes a general analysis of the proteins expressed and assessment of the specific properties of the GM food under consideration to elicit hypersensitivity reactions. A better understanding of the impact and interaction of food with the immune system is required to decipher how and whether conventional and GM foods cause specific health and safety problems.

New methodology for the development of GMOs may significantly reduce potential risks derived from the random integration of transgenes used in current methods.

Developing Regulatory and Safety Systems for Modern Food Biotechnology: A Role for Capacity Building

The United Nations and international development agencies coined the term 'capacity building' in the early 1990s after an appraisal of the development-assistance programmes in developing countries. Within that context, the term has come to mean different things to different people. In 1997, a World Bank progress report on Africa defined it as: "...an investment in people, institutions and practices that will, together, enable countries in the region to achieve their development objectives" (World Bank 1997). Capacity building is a four-step process involving a needs assessment, strategic planning to change the situation, training of personnel to implement the changes, and an evaluation of the results.

A United Nations Development Programme (UNDP) report describes capacity building as a continuous process that ought to take place at various levels: individual, institutional and societal. The first two levels involve expanding local knowledge and skills. At the societal level, it is about creating opportunities to engage the trained individuals to their fullest potential. All three levels are interdependent and need to be pursued concurrently in order to achieve the maximum benefit.

In a report published in 2000, UNDP acknowledged that due to the different levels of development among countries, some countries may never be in a position to deploy cutting-edge technologies. Nevertheless, these countries need local expertise to understand and adapt technologies for national use, consistent with their development goals.

It is widely accepted that the application of modern biotechnology could be important to economic development, but may also involve inherent risks (UNECA 2002; CBD 2005d). Thus, all countries, be they developers or net importers of products derived from modern biotechnology, should introduce measures that safeguard human health and environmental safety. In fact, many governments are in the process of developing legal instruments/regulatory systems that address human health and environmental safety. The effectiveness of such measures will be determined by a country's capacity (both in terms of human resources and infrastructure) to expeditiously handle the evaluation, management and risk communication of each new product of modern biotechnology. While evaluation and risk management may be done on a case-by-case basis, risk communication activities undertaken by governments should address the process according to which decisions are taken.

Starting in the 1970s, R&D in biotechnology has been important to development collaboration (Jenny 1999). This trend was supported by the adoption in 1992 of Agenda 21 (UNDESA 1992) and, more recently, the Convention on Biological Diversity. These two agreements include specific sections on the application and use of biotechnology in major economic sectors such as agriculture, industry and energy. To complement this, many donor agencies, NGOs, the private sector and governments in industrialized countries have focused their capacity-building policies and objectives on maximizing the benefits of biotechnology in developing countries through technology transfer/extension. The database of capacity-building initiatives of the CBD Biosafety Clearing House (CBD 2005b) demonstrates the strategies intended to target progress towards the Millennium Development Goals (World Bank 2000b).

Agenda 21 (UNDESA 1992) is a comprehensive plan of action to be taken globally, nationally and locally by organizations of the United Nations system, governments and major groups in every area in which humans place a burden on the environment. It was adopted in June 1992 by over 178 governments at the United Nations Conference on Environment and Development (UNCED 1992).

This approach has addressed one specific area (technology development), but has failed to impart the skills and knowledge necessary to undertake associated activities, such as development and implementation of regulatory and food-safety frameworks.

Safety issues with regard to protecting the environment and human health are different and require different expertise. Biosafety tends to be the responsibility of the department of environment or agriculture, whereas the authority for food safety often lies with the department of health. Hence, the legal instruments for regulation may differ.

At the international level, 15 legally binding instruments and non-binding codes of practice address some aspect of GMOs, but none of these on its own integrates the regulation of biotechnology across all sectors (Glowka 2003). Such sector-based regulations and powers increase the already overstretched capacity needs of developing countries, and present challenges to developing a fully coherent policy and regulatory framework for modern biotechnology (FAO 2002). The challenge for developing countries is to achieve coherence in national legislation for crops, livestock, fish, forest trees and microorganisms, while meeting international obligations and ensuring harmonization (Glowka 2003).

The shortcomings of most capacity-building programmes lie in the simplistic notion that assumes a 'one size fits all' development path (Fukuda-Parr et al. 2002). Donors often prescribe programmes that are largely based on the experiences of developed countries, on the assumption that these will work equally well for developing countries. Unfortunately, this is rarely the case and can result in limited or disappointing outcomes.

A sound capacity-building programme is determined by its ability to focus on human development, in order to foster the skills and resources needed to sustain its own progress. In other words, a capacity-building initiative must act as a support and catalyst to self-reliance and tap into a country's ability to master its own development, in harmony with its natural environment and any other national imperatives such as economic sustainability (ECDPM 2003).

Capacity-building initiatives must be sustained beyond the life of the activity as an integral part of a development programme and not be a once-off activity (Anon. 1999). In turn, developing countries must participate in and take ownership of an activity and be encouraged to take charge of their own development. Demand-driven knowledge development is more likely to be absorbed if it reflects local circumstances, and more likely to be applied by society.

Capacity Needs

Food safety is attracting increased attention because of its implications for public health (World Bank 2000a). In general, food control systems in developing countries are poorly developed, and less organized than in most developed countries. Their overall capacity needs in terms of food safety can be summarized as follows (FAO 1999a): (1) basic infrastructure; (2) national food control strategy; (3) food legislation and regulatory framework; (4) food inspection services; (5) food control laboratories and equipment; and (6) implementation of food quality and safety assurance systems.

The work in food safety is multidimensional, and there are frequently several food laws under the authority of different agencies (WHO 2002b). In many countries, effective food control is undermined by the existence of fragmented legislation, multiple jurisdictions and weaknesses in surveillance, monitoring and enforcement mechanisms. Food-safety legislation developed specifically for the safety of GM foods should be integrated within the existing food laws, taking into account the special risk-management requirements.

In order to make informed decisions on the safety of GMOs and GM foods, governments need substantial human and institutional resources in the disciplines required for assessing the risks for the environment and for human food presented by GMOs. Developing countries have limited expertise in the required fields of science, as biotechnologists in these countries are generally engaged in research and therefore mostly unavailable to the regulatory bodies and as policy-makers (Mugabe 2000). In most developing countries, those same scientists sit on national biosafety committees, and are involved in both risk assessment and policy-making. There are three vulnerabilities in this scenario: (a) when developers are also risk assessors, the potential for conflict of interest is magnified; (b) because most members of the national biosafety committee are recruited on a voluntary basis, they do not devote too much time to this responsibility; and (c) because membership of the national biosafety committee generally rotates, there is no continuity in the capacity gained through experience. While many developed countries have adopted mechanisms to govern modern biotechnology, most developing countries are either in the process of developing national biosafety frameworks or are yet to start the process. To date, no more than 10 developing countries have implemented national biosafety laws (CBD 2005c). A further 20–30 are in a state of transition whereby some or all elements are at different stages of development. A few developing countries that permit the commercial planting of crops derived from modern biotechnology have modest capacities to implement a regulatory framework (Paarlberg 2001b).

Where national biosafety frameworks are in place, they vary between countries according to national priorities and statutory structures. In addition, the different social conditions that prevail in different countries make it difficult to determine the appropriate regulatory systems that should be enforced by developing countries (Nuffield Council on Bioethics 2003). Notwithstanding the diversity, a number of elements are essential and form the core of many national frameworks:

- national policy and strategy;
- regulatory framework consisting of regulations and guidelines;
- mechanism for handling applications and issuing permits;
- system for enforcement; and
- system for information dissemination.

The impetus to establish regulatory frameworks for biosafety seems to be a significant factor in determining the process whereby they are developed. In some cases, scientists have raised interest in regulating local research, while in others the trigger may have come from multinational companies seeking to continue seed production in the Southern Hemisphere during the northern winter months. Recently, the importation of food aid has triggered some form of regulation in those countries that have been faced with food shortages.

Many countries with regulatory systems have developed and implemented these systems in a stepwise fashion, usually in response to an immediate demand (Cohen 2001). The first step has involved the establishment of voluntary guidelines to set in process a structured progression of the regulatory framework. The guidelines initially set the principles for safety in laboratory practices, which are later adapted to ensure environmental safety for enabling field trials.

The advantage of guidelines is that revising and incorporating new information requirements in line with an evolving technology can be done very swiftly. However, guidelines are voluntary and compliance cannot be enforced unless supported by regulations (McLean et al. 2002).

Institutional and Human-Resource Constraints

Many countries face major constraints with respect to enhancing their regulatory capacity needs. These constraints fall into three categories: institutional, human resources, and cost (Juma and Konde 2002). The first two are interdependent in many respects.

The Codex *Principles for the risk analysis of foods derived from modern biotechnology*, adopted in 2003 (CAC 2003b), recognize the need for improving the capabilities of regulatory authorities in handling risk analysis. Capacity-building programmes for developing countries are also being discussed within the Codex system.

Capacity building is one of the essential elements of the Cartagena Protocol on Biosafety (CBD 2000). Its *Article 22* is devoted entirely to this issue, while paragraph *3* of *Article 28* deals with the financial support that developing countries may require in meeting their capacity needs if they are to be effective in implementing the Protocol.

Countries with a weak knowledge-and skills-base tend to develop highly protective regulations at the expense of innovation. In contrast, flexibility in regulatory structures tends to be encouraged by a broad knowledge and capacity base (McLean et al. 2002).

The capacity-building needs of developing countries can be grouped according to (among other aspects): the level of biotechnology research; the capabilities to develop marketable products; the level of development that would determine whether a country becomes an importer or exporter of products derived from modern biotechnology. This last issue is of crucial importance in the needs assessment. It enables a country with limited resources to plan and invest realistically in the capacities that will be used.

Financial Constraints

Recognizing the need to regulate modern biotechnology, and appreciating that developing countries may need to re-evaluate their spending priorities, the cost implications of establishing national biosafety regulatory frameworks, including GM food-safety regulations, should be assessed. A country's financial situation has an overriding influence on the development and implementation of national frameworks for the regulation of modern biotechnology. For a framework to be effective, an identification of existing resources, gaps and training should be made in order to build on the expertise and experience available in a given country. However, the national priorities of developing countries may differ from those of developed countries, so that these governments may elect to use their limited resources in other ways.

Cost in itself raises important questions with respect to finding the right balance between meeting obligations under international agreements and addressing national priorities. The cost of establishing a national biosafety framework, including a food-safety framework, will vary dramatically among countries according to their judicial systems, their individual capacities, and their regulatory objectives.

In 2002, the World Bank and the World Trade Organization (WTO) announced the launch of a fund, the Standards and Trade Development Facility (STDF), in collaboration with FAO, the World Organisation for Animal Health (Office International des Epizooties; OIE) and WHO. The main objective of the Facility is to coordinate the activities of the international organizations in order to maximize the financial and technical support given to developing countries for implementing international standards for food safety, plant and animal health.

In 2000, the Council of the Global Environment Facility (GEF) agreed to support the *Initial strategy for assisting countries to prepare for the entry into force of the Cartagena Protocol on Biosafety*, a three-year project implemented by the United Nations Environment Programme (UNEP), for the establishment of country-driven national biosafety frameworks. This project, initiated in June 2001, had enlisted 123 countries globally as of September 2004 to set up frameworks for the management of products derived from modern biotechnology at the national level, and it is hoped to establish cooperation at subregional, regional and international levels.

According to UNEP and GEF, 139 countries meet the criteria set and therefore qualify to participate in the *Development of national biosafety frameworks project* estimated at US$38.4 million. Assuming the enlisted 123 countries get assistance from the project, an estimated US$400,000 per country is required to establish a national framework. One-third of this total is to be contributed by the country in cash or in kind.

The MATRA project of the Netherlands Ministry of Foreign Affairs invested US$60,000 per country in the pre-accession countries of Central and Eastern Europe (CEE). These funds were used over a three-year period to establish national biosafety frameworks that conform to the relevant European Community directives and the CPB.

When the project was initiated, the CEE countries were at different stages of developing national frameworks as some (e.g. Hungary and Poland) had benefited from the UNEP/GEF *Pilot biosafety enabling activity project*. At its conclusion, the countries were not only ready to join the EU, but had a regional web site hosting information on their frameworks and regional activities, and established centres of excellence that will sustain capacity development in the region.

Other costs that need to be taken into consideration include systems for monitoring compliance, and costs associated with review of the scope and effectiveness of the legal requirements in keeping pace with new scientific developments and public opinion.

Food Safety Capacity Development

In order to support countries wishing to fulfil the mandate set by the CPB (CBD 2000), the secretariat of the CBD maintains a global database of capacity-building initiatives as a component of the Biosafety Clearing House (CBD 2005b). The purpose of this database is to give an overview of past, present and prospective capacity-building initiatives. The secretariat intends to use the information to develop a method for coordinating capacity-building initiatives, thus ensuring that they complement one another, use funding efficiently and strengthen resources in the recipient countries. Although the secretariat's interest is in initiatives that would support the effective implementation of the CPB, the database covers a wider range of initiatives, such as technology transfer and those directed towards biotechnology research.

To date, 89 initiatives have been listed in the database, illustrating a broad range of implementing agencies. According to the secretariat, more than half of the registered initiatives have been negotiated bilaterally and through industry interest groups. United Nations agencies, intergovernmental organizations, or individual governments, industry or NGOs supported most of these countries through bilateral agreements. While the capacity-building initiatives collectively cover all the aspects associated with the application of modern biotechnology, no single one covers the entire range — each is limited to its own specific focus. For example, the FAO/WHO expert consultations and capacity-building programmes supported by both organizations train individuals in food-safety-related issues only.

WHO has advised Member States and assisted in building their capacity for food-safety-related issues for many years. The food-safety activities of WHO have increased significantly over the years, with the establishment of international expert scientific bodies such as the Joint FAO/WHO Expert Committee on Food Additives (JECFA) in 1956, to evaluate the safety of food additives, contaminants, naturally occurring toxicants and residues of veterinary drugs in food; the Joint FAO/WHO Meeting on Pesticide Residues (JMPR) (1963) to evaluate the safety of pesticide residues in food; the Joint FAO/WHO Expert Meeting on Microbiological Risk Assessment (JEMRA) (2000) to provide risk-assessment guidelines for selected pathogens and for microbiological hazards in food and water. Also in 1963, the Codex Alimentarius Commission was created to implement the joint FAO/WHO Food Standards Programme.

To strengthen its in-house activities, WHO created the Food Safety Programme in 1978, operating at national, regional and international levels. The recognition of food safety as a major public-health concern by the World Health Assembly in 2000 has also increased the profile of food-safety-related issues, not only within the Organization but also at the national level (WHO 2000b). These activities were further supported by the endorsement of the WHO Global Strategy for Food Safety by the WHO Executive Board in 2002.

In this strategy, WHO proposes to "formulate regional food safety strategies on the basis of the WHO global food safety strategy and of specific regional needs such as technical support, educational tools and training".

Considerable technical assistance has been provided to developing countries to create and/or enhance food-safety control systems, but these activities have not been effectively coordinated and therefore not been adequate in meeting the public-health needs of recipient countries.

The SPS Agreement of WTO (WTO 1995, Article 9) calls for assistance to developing-country Members to enable them to strengthen their food safety and animal and plant health protection. The

Agreement encourages Members to enter into bilateral arrangements for technical assistance, or to seek training through other international organizations. Such assistance can be in the area of processing technology, research or infrastructure development, and may take the form of technical advice, expertise, financial assistance or procurement of adequate equipment.

As previously mentioned, food-safety activities within WHO take place at the international, regional and country levels. The regional and country offices provide assistance in developing and strengthening national food-safety programmes, whereas WHO headquarters develops guidelines for such work, including the framework for risk analyses and setting international standards (Mahoney 2001). The division of these activities is arbitrary as headquarters also participates in activities at the national and regional levels, with technical know-how and capacity-building guidance. These activities include (FAO/WHO 2003b):

- developing regional and national food-safety policy and strategies;
- preparing food legislation, regulations, standards and codes of hygienic practice;
- implementing food inspection programmes;
- promoting methods and technologies designed to prevent foodborne diseases, including the hazard analysis and critical control point (HACCP) system;
- developing or enhancing food analysis capability;
- developing methods for assessing the safety of the products of new technologies;
- establishing healthy markets and enhancing the safety of street food; and
- promoting the establishment of foodborne disease surveillance systems.

Many WHO activities to build food-safety capacity are developed in collaboration with FAO. However, FAO also administers a major, separate technical cooperation programme building capacity in this and other agriculture-related areas in many developing countries.

Although most developing countries have national food-control systems, these are often not based on modern scientific concepts.

Moreover, they cannot be adapted to cope with developments in food science and technology (Gupta 2002). The specifications for an effective food-control system include: regulations, capacity for assessing the risks associated with the food, and ongoing monitoring and evaluation of the risks. A capacity-building programme for the risk assessment of products of modern biotechnology would involve:

- use of the concept of a comparative safety assessment; hazard identification and characterization;
- assessment of food intake, including consumption profile and effects;
- use of integrated toxicological evaluation;
- use of integrated nutritional evaluation;
- risk characterization; and
- application of risk-management strategies, such as labelling and monitoring.

Other Considerations

Apart from the human resources and physical facilities in which to perform biosafety-related research, competent authorities need information relating to trends in biotechnology and biosafety to keep abreast with biotechnology developers. Information exchange systems as provided by a number of organizations fulfil this need by facilitating international cooperation, but can only be used by developing countries where the appropriate expertise exists. Even more limiting, several of these information networks are difficult to search, while others are limited in scope (Louwaars et al. 2002). The Intergovernmental Committee for the CPB, realizing the capacity constraints of developing countries, has established a coordination mechanism to maximize synergies, complementarity and collaboration between the numerous international initiatives.

The evaluation of food is not only about science. It should also take into account the social, ethical and religious concerns of the local populations.

Harmonization

At the international level, protocols have been agreed upon that implicitly promote the harmonization of regulatory systems. While the Codex *Principles for the risk analysis of foods derived from modern biotechnology* (CAC 2003b) are available to guide the safety assessment of GM food, they have no binding effect on national legislation, but do form the basis for harmonization under the SPS Agreement (WTO 1995, Article 3.4). On the other hand, the CPB has established legally binding rules for environmental risk assessments (CBD 2000). In addition, OECD has experience in promoting international harmonization in the regulation of biotechnology by ensuring efficiency in the evaluation of environmental and human-health safety, through its working group for harmonization in biotechnology and its task force for the safety of novel foods and feeds (OECD 1995, 1996).

Developing countries therefore have sets of agreed principles (regulatory and risk assessment of foods) for guidance, and the advantage of learning from the experiences of their forerunners by investigating best practices and adapting them to suit their individual situations.

Although agreement has been reached on the scientific principles of food-safety assessment, consensus has not been achieved on the extent of data required to comply with these principles or on the role of the data in decision-making.

Harmonization of components of the scientific review process has a potential benefit where a lack of resources threatens effectiveness, and the affected countries in the region have determined and agreed on the regulatory objectives. The advantages of regional/subregional cooperation are to facilitate regulation, promote the sharing of resources, synchronize the assessment of foods derived from modern biotechnology, and expedite information exchange (McLean et al. 2002). The Nuffield Council of Bioethics (2003) recommends the implementation of international standards and the sharing of risk-assessment methodologies and results, particularly between developing countries with similar ecological environments.

Moreover, integrating some activities could reduce the overall requirement for new financial resources. Harmonization can be achieved at several levels, i.e. some elements of the framework can be implemented at the regional level. The countries of the Association of South-East Asian Nations (ASEAN) have come together to cooperate on various levels, including: (i) harmonization of legislation for products derived from modern biotechnology and intellectual property rights; (ii) R&D in biotechnology; and (iii) environmental protection. ASEAN is also looking at a regional approach to biosafety, although it is not clear what is intended, i.e. whether regional assessment and national decision-making would be considered. Those countries in the region that have made some progress have gone as far as developing labelling regulations, although they acknowledge that implementation may not be possible in the near future due to a lack of human resources.

After the 2002 humanitarian crisis in southern Africa, where a number of countries experiencing severe drought and food shortages questioned the use and safety of GM food aid, a Council of Ministers of the Southern African Development Community (SADC) established an Advisory Committee on Biosafety and Biotechnology (SADC 2003) to develop a common position on biotechnology and harmonize biosafety legislation in the region. The objective is to facilitate the movement of food products that may contain GM material across the region in future.

Although harmonization may absorb some of the costs that could be incurred in establishing regulatory frameworks, the flexibility allowed by international agreements creates room for divergence from the basic principles. Also, none of the regimes give guidance on regulations. Therefore, achieving harmonization in this context may be debatable as countries grapple with criteria for the precautionary approach and socioeconomics. Nevertheless, particular attention should be paid to supporting and creating new strategic partnerships. Countries need to find effective ways of working together, and analyse the benefits and costs of harmonization.

Many capacity-building initiatives to date have tended to address a specific need: to develop competency for implementing an international treaty. Nonetheless, several are independent and not linked to any international treaty.

The broad information base required for decision-making with regard to the adoption of modern biotechnology indicates that developing countries need a clear understanding of all the issues. To develop this awareness, the development of human resources needs to extend beyond biosafety training and include food safety, intellectual property rights management, and trade issues. The relevant intergovernmental organizations (CBD, FAO, UNEP, WHO and WTO) should consider coordinating their capacity-building efforts to achieve this holistic approach to imparting knowledge and supporting national capacity building.

Many developing countries cannot afford the seemingly considerable capacities required for the adoption of modern biotechnology. Measures must be taken to ensure that developing countries are not impeded in effective regulation by development problems, and that they derive benefit from their participation in international regulatory instruments.

One way of safeguarding developing-country interests would be to establish a global roster of experts, ideally with a regional balance. However, experience in biosafety is largely gained on the job. Therefore, scientists that may have had exposure to international discussions or even had training may not necessarily know which questions to ask in a safety assessment, because their training may have been theoretical and not given them experience of a real-life situation.

Alongside the above-mentioned activities, a potential normative role exists for WHO to coordinate the scientific food-safety assessments of products of global importance.

A Potential Role for Modern Biotechnology

The Convention on Biological Diversity dictates the use and application of relevant technologies as a means of achieving the objectives of conservation and sustainable use with specific reference to biotechnology.

Modern biotechnology is purported, from a technical perspective, to have a number of products for addressing certain food-security problems of developing countries. It offers the possibility of an agricultural system that is more reliant on biological processes rather than chemical applications (Rosegrant and Cline 2003). The potential uses of modern biotechnology in agriculture include: increasing yields while reducing inputs of fertilizers, herbicides and insecticides; conferring drought or salt tolerance on crop plants; increasing shelf-life; reducing postharvest losses; increasing the nutrient content of produce; and delivering vaccines (Bonny 1999). The availability of such products could not only have an important role in reducing hunger and increasing food security, but also have the potential to address some of the health problems of the developing world.

Achieving the improvements in crop yields expected in developing countries can help to alleviate poverty: directly by increasing the household incomes of small farmers who adopt these technologies; and indirectly, through spill-overs, as evidenced in the price slumps of herbicides and insecticides. Indirect benefits as a whole tend to have an impact on both technology adopters and non-adopters, the rural and urban poor. Indeed, some developing countries have identified priority areas such as tolerances to alkaline earth metals, drought and soil salinity, disease resistance, crop yields and nutritionally enhanced crops.

The adoption of technologies designed to prolong shelf-life could be valuable in helping to reduce postharvest losses in regionally important crops. Prime candidates in terms of crops of choice for development are the so-called 'orphan crops', such as cassava, sweet potato, millet, sorghum and yam. Multinationals have found no incentive to develop these crops and have instead invested in marketable crops with high profit returns. This strategy is intended to target wealthier farmers in temperate-zone countries with the financial capacity and tradition of supporting new seed products. However, here is a potential for multinational companies to develop crops grown largely in developing countries. The investment costs are low and the potential markets considerably large (Conway 1999).

While some public-sector research institutes in developing countries are forging ahead with the application of modern biotechnology, a small number are supported by government policy and therefore follow a defined agenda (Skerritt 2000). Still other governments believe that the risks (safety, environmental and/or economic) associated with modern biotechnology outweigh the benefits.

Currently, the many promises of modern biotechnology that could have an impact on food security have not been realized in most developing countries (Luijben and Cohen 2000). In fact, the uptake of modern biotechnology has been remarkably low owing to the number of factors that underpin food security issues. In part, this could be because the first generation of commercially available crops using modern biotechnology were modified with single genes to impart agronomic properties with traits for pest and weed control, and not complex characteristics that would modify the growth of crops in harsh conditions. Secondly, the technologies are developed by companies in industrialized countries with little or no direct investment in, and which derive little economic benefit from, developing countries. Thirdly, many developing countries do not have the necessary biosafety frameworks to regulate the products of modern biotechnology. For example, it took over two years for the Kenyan authorities to approve the field-testing of a virus-resistant sweet-potato variety because the scientific capacity for evaluating the product was unavailable (Juma

2001). It should be noted, however, that such delays in the approval process have also been seen in developed countries, especially during the initiation of national regulatory evaluation.

However, this trend is quickly changing as a number of developing countries either adopt or develop appropriate biotechnologies or regulatory infrastructures. A report by the International Service for National Agricultural Research states that more than 40 crops are the focus of public-sector research programmes of 15 developing countries involving disease-resistant traits in rice, potato, maize, soybean, tomato, banana, papaya, sugarcane, alfalfa and plantain (Paarlberg 2001a). Some of the local crops in the priority research list of research institutions in developing countries. For example, the Brazilian Agricultural Research Corporation has concentrated research into genetically modified crops on disease resistance in beans, papaya and potatoes. The research programme at the University of Cape Town (South Africa) focuses on the development of crops resistant to viruses and to desiccation. The university has recently had a breakthrough with maize streak-virus resistance. In Thailand, the National Centre for Genetic Engineering and Biotechnology has supported research into disease resistance of rice, pepper and yard-long beans.

Although current commercial GM crops are not designed to address the specific issues of developing countries, their adoption has shown that they can be relevant in some developing countries — for example, the planting of herbicide-tolerant soybeans in Argentina and *Bt* cotton as a cash crop by resource-poor farmers in China and South Africa (Paarlberg 2001a).

There is little information on the economic costs associated with R&D of products of modern biotechnology, or on the impact of their introduction on production costs. An in-depth analysis of the short- and long-term economic and social costs and benefits is necessary.

Qaim and Zilberman (2003) report that farmers in Argentina that adopted herbicide-tolerant soybeans reduced per hectare production costs through the reduced number of herbicide applications, and thereby increased total factor productivity by 10%.

On average, the *Bt* cotton farmers in China reduced pesticide spraying for the Asian bollworm by 70%, producing a kilogram of cotton at 28% less cost than the non-*Bt* farmers (Huang et al. 2002b). These benefits have had a significant impact on the agronomic, environmental, health and economic situations of approximately 5 million resource-poor farmers over eight provinces. Similarly, farm-scale trials in China of GM rice containing genes which make them resistant to insect larvae that devastate rice crops showed 80% less pesticide use and yields increased by 6–9% (Coghlan 2005). In addition, farmers who grew the GM varieties suffered less pesticide-induced illness than those growing the old varieties (Coghlan 2005).

A two-year study of the economic impact of *Bt* cotton adoption by the farmers in the Makhathini Flats of Kwa-Zulu Natal Province of South Africa showed that farmers not only experienced yield increases, but that the savings from reduced chemical applications outweighed the higher seed cost (Ismael et al. 2001). Between 1997 and 2001, the number of South African cotton farmers who adopted the planting of *Bt* cotton increased 16-fold (Bennett et al. 2003).

Several agro-economic studies have been commissioned since the introduction of seed derived from modern biotechnology in the USA. One report illustrates that the greatest yield increases were achieved with insect resistant maize, while the greatest reduction of input costs was seen in herbicide-tolerant soybean (Gianessi et al. 2002). The economic benefits associated with the cultivation of *Bt* maize by farmers in the USA in 2001 were primarily the result of the decreased need for pesticides. The financial gain takes into account the seed-price premium paid by farmers for *Bt* maize seed. Benbrook (2002) argues that farmers in the maize belt forfeit a significant proportion of their farm income to biotechnology companies because of the seed-price premium.

While evidence shows that GM crops can lead to significant productivity and health gains, they are nevertheless not a 'magic bullet' that will solve all problems in agriculture. Modern biotechnology must be applied to complement and expand the reach of conventional methods (Pingali 2001). It has been alleged that focusing on modern biotechnology may narrow the research agenda of many countries and deny them the opportunity to explore solutions that can be freely adopted, adapted and exchanged (UNECA 2002). For instance, where the cause of declining farm productivity can be attributed to poor soil fertility, the current technologies do not provide any remedies. On the other hand, almost half of the world's potentially cultivable tropical land has acidic soil, caused by excessive soil aluminium (Herren 1999). The production of GM aluminium-tolerant crops would allow the productive cultivation of millions of hectares of acidic soil lands in tropical Asia and Latin America (Herrera-Estrella 1999). It should also be borne in mind that conventional breeding is still the technique most often used for achieving yield increases and for developing crops with resistance to diseases, insects and abiotic stresses (de la Fuente et al. 1997). Moreover, conventional breeding still contributes the bulk of new crop varieties used in general. It is, however, alleged that with the anticipated increase in the world population over the next 25 years, grain production will need to increase by 26 million tonnes per year. In addition to traditional breeding methods, it may be necessary to apply other techniques to achieve the required yield increases and yield stability of rice and other grains (Huang et al. 2002a).

Developing countries with limited financial and human resources need to find the right balance for investing in conventional and modern biotechnology research programmes. While alliances with the private sector may contribute to the search for new technologies, the public sector needs to focus on crops and traits in which the former may be unwilling or unable to invest (Khush 2003). The extent to which priority is given to modern biotechnology over other research methods should be linked to a country's agricultural priorities and objectives as well as to its environmental concerns.

Ultimately, investment in interventions that support good governance, the development of rural infrastructure and market access is required before any of the promises of modern biotechnology can be realized. In general, policies that stimulate economic growth and target poverty reduction may have significant bearing on the health and well-being of the population (Luijben and Cohen 2000).

Research Ownership

Research is a critical part of any effort aimed at improving food production and reducing poverty. Globally, much of agricultural R&D is carried out in the public sector, thereby serving the interests of developing countries (Conway 1999). Public research in developed countries and Latin America is mostly conducted by government institutions and universities, whereas almost all agricultural research in Africa is carried out in public institutions, including R&D, technology transfer and dissemination of improved plant varieties (Cohen and Pinstrup-Andersen 2002). In general, international agricultural research institutions form a second level of research development and technology providers in developing countries. Public research institutes have, in the past, researched and improved orphan crops, mainly for donation to poor farmers or at cost.

Generally, academic institutions are perceived as producers of knowledge that benefit and protect the public. Also, national and international research institutes aim to address the agricultural problems of resource-poor farmers in developing countries, e.g. increasing productivity through the use of a variety of techniques, including modern biotechnology (Pardey et al. 2001a). In reality, public institutions are now exposed to the forces of globalization and compelled to compete for their survival.

In the current climate, government intervention in R&D worldwide has dwindled; hampering the level of innovation generated for public good. In fact, research facilities in many developing countries are poorly equipped, often limiting experiments to traditional and outdated research. The diminishing role of public research institutes is perceived to have a major impact on the adoption of modern biotechnology in terms of introducing relevant products to those that need them most.

Most of the field trials in the EU and the USA are conducted by private companies (Fresco 2003). An analysis of field-trial data from the USA shows that three crops (maize, potato and soybean) account for 64% of all trials, of which 69% express herbicide-and pest-resistance traits. Of the trials conducted in the EU, 67% involve maize, sugar beet and rapeseed, and 71% of the novel gene categories presented herbicide-or pest-resistance traits. Less than 1% of all the trials in the EU and the USA are of plant varieties grown in tropical and subtropical climates, half of which have been conducted by the public sector.

Most of the public-sector research involving modern biotechnology in developing countries (except China) is still in the laboratory phase — none of the crops have progressed to marketable products (Arundel 2002; Fresco 2003). China, on the other hand, has approved the field-testing of over 500 GMOs to date and the commercial release of 50, including a prolonged shelf-life tomato, virus-resistant sweet pepper and vaccines for animal use.

The experiences limiting the progression to commercializing research efforts range from: a lack of resources for meeting the high costs of regulatory requirements; lack of foresight, planning and business acumen for enabling the transition from research to a commercial product; and lack of capacity to negotiate patent licenses.

Also, developments in modern biotechnology have occurred independently of the sustainable agricultural goals and priorities of the developing countries concerned. Furthermore, a needs assessment for a particular technology has often not been carried out before a research project is begun (Taylor and Fauquet 2000). Nevertheless, it is often argued that the commercialization of some products would encourage monocropping as national agricultural research has focused on a few crops, whereas rural communities tend to grow a wide range of crop species and plant varieties. The focus of international agricultural research centres is on plant production and protection (78%), livestock production and health (21%) and food processing (1%). With respect to food crops, research emphasis appears to be spread equally between cereals, root crops and legumes.

However, within the cereals, research devoted to rice far outweighs research on maize and sorghum (Taylor and Fauquet 2000). A large proportion of the total agricultural research activities of many developing countries are donor-funded. International research institutes, such as the Consultative Group on International Agricultural Research, depend on government grants and donations from philanthropic organizations for their survival, and yet investment in this sector has fallen in real terms (Pardey et al. 2001a). A significant fraction of the funding spent on international agricultural research institutes is used on activities covering a relatively large number of crops. The beneficiaries of such initiatives are a small number of countries with relatively advanced scientific capabilities (Pardey et al. 2001a).

During the 1990s, developing countries as a group invested more in agricultural research than developed countries, even though the spending was unevenly distributed. In industrialized countries, private-sector investment in R&D far exceeds government spending on technology development, so that much of the public good previously entrusted to public research institutes is now privately owned. In comparison, private-sector investment in developing countries is around 1% of total global spending in this sector (Taylor and Fauquet 2000) and developing countries invest less than 5% of the total private sector spending in biotechnology. Although the agricultural sector in developing countries is large and of significant importance to the domestic economy, spending in agricultural research does not match this level of activity. For instance, 80% of the food consumed in sub-Saharan Africa is obtained from domestic production (FAO 2002b).

Impact of Intellectual Property Rights on Research

Intellectual property rights (IPRs) have been relevant to agriculture since their inception but have gained importance with regard to research in developed countries in the past 20–30 years. In particular, IPRs have been used to protect and preserve the value of products produced by conventional methods, e.g. the trademark registration of food products (Dutfield 2001). The rationale for IPRs is that they encourage the inventor to advertise the invention and disclose the new knowledge, while simultaneously holding the rights to protect the invention from competitors (Taubman 2004). Accordingly, disseminating this information is thought to stimulate new ideas and further rounds of innovation and technological advancement. IPRs afford time-limited protection to artistic, scientific, technological or economic products, and can be protected by way of copyrights, trademarks, design patents, utility patents, plant patents, plant breeders' rights and trade-secret law. Of these mechanisms, patents are considered the most powerful tool of the IPR system (Wendt and Izquierdo 2001).

Patents play different roles in different technologies and sectors. Patent protection of biotechnology makes it a tool for technology transfer and securing new markets in a global economy (Barton and Berger 2001). Without protection, new ideas and information are entirely in the public domain. This can, in certain systems, lead to underinvestment in R&D or the withholding of knowledge (Wendt and Izquierdo 2001).

Plant variety protection (PVP) provides less protection than patents in that generally it makes provision for farmers' rights, allowing them to use harvested seed, and includes an exemption for research use. Despite the increase in availability, new plant varieties continue to be inaccessible or inappropriate for poor farmers, and the rate of innovation remains largely unchanged in countries with a PVP system (Pardey et al. 2001b). Studies have indeed shown that in middle-income countries, the principal beneficiaries of PVPs are commercial farmers and the seed industry.

PVP is seen as a system that protects small advances in plant breeding, while a patent regime is thought to lead to the protection of big leaps in technological achievements (Helfer 2002). Patent protection for products of modern biotechnology is important because they are expensive to develop and easy to copy. Even so, developing countries have limited capabilities to innovate in industrial fields such as modern biotechnology, and to effectively enforce IPRs. A significant number of developing countries have not established intellectual property regimes that cover plants. This situation may thus discourage private-sector investment (Chaturvedi 2001). With no assurance that they can recoup some profit on GM products, multinationals are unlikely to devote much attention to the challenges of developing countries unless seen in a development aid context or through public–private partnership. Although this situation impedes private-sector investment in developing countries, it also implies that the freedom to operate on products designated for local markets is not hindered (Wendt and Izquierdo 2001).

Exercising this freedom to operate is not a well-understood concept. For example, in the case of 'golden rice', where permission was required for the use of about 70 patents, the impression was that the patents were being relinquished in favour of the poor. In fact, most of the patents involved are not valid in the major rice-consuming countries. The technology donated for the development of virus-resistance in non-commercial potato varieties is free of patents relevant to Mexico, and the same holds true in the case of virus-resistant sweet potatoes in Kenya. Researchers are usually unaware of the proprietary status of the technologies they are using in their work.

Nevertheless, the proliferation of broad patents is thought to impede the research capabilities of other interested parties (Salazar et al. 2000). Some countries grant very broad patents conferring monopoly rights over large areas of research, thereby potentially threatening the other goal of intellectual property, namely the right to build upon the original invention. The prevailing patent rules have the potential to limit the accessibility of these technologies to public institutions and ultimately poor farmers (Krattiger 2002). Furthermore, the strengthening of IPRs is thought to restrict the flow of germplasm and inhibit the development of new plant varieties (Barton 1999). This is because if and when researchers in public institutions do get permission to develop the technologies further, access is granted under licence agreements with restrictions on commercializing innovations. It is also argued that a stringent, multilateral IPR system will not benefit all countries equally. Indeed, the benefits will largely be influenced by the economic and technological levels of development in each country (Barton and Berger 2001).

According to the United Kingdom Commission on Intellectual Property Rights, "the critical issue in respect of IPRs is perhaps not whether it promotes trade or foreign investment, but how it helps or hinders developing countries to gain access to technologies that are required for their development."

During the 1990s, obtaining a patent application (excluding the cost of filing, which varied in different countries from US$355–4771) in the USA cost US$20,000 and twice as much in the EU (Barton and Berger 2001). In general, PVP is cheaper; valued at one-tenth of the patent price. Moreover, the preparation of a food-safety dossier for a product derived from modern biotechnology, for example, is estimated to be around US$1 million (Tansey 1999). These estimates cannot be compared with the regulatory costs in developing countries. Although significantly lower than figures quoted above, the cost of regulation in developing countries does not encourage the commercialization of products of modern biotechnology developed by public-sector research institutes (Lesser 1997). In most cases, the regulatory costs far exceed the research costs.

Many developing countries do not have the resources to match private-sector investment in modern biotechnology. In this new playing field, public institutions also need resources to deal with intellectual property rights to help compensate and increase the public benefit. Otherwise their involvement in R&D could be deterred by lack of funding. If public institutions are to use the techniques of modern biotechnology, then the use of IPRs as a framework for facilitating technology transfer must be emphasized more than its handling as a revenue-generating system. IPRs can, however, play a major role in clarifying the mechanisms for access to technology, and determine the downstream aspects of use and exploitation of genetic resources.

There are several ways in which public institutions and small companies in developing countries can gain access to patented genes and enabling technologies to overcome the current barriers to research. The first of these includes a measure of goodwill by multinationals to relinquish their rights to technologies for use by researchers in developing countries by adopting programmes of social responsibility as in the case of 'golden rice', a variety containing beta-carotene (a precursor of vitamin A), virus-resistant sweet potatoes (in Kenya) and a virus-resistant non-commercial potato variety in Mexico (Falck-Zepeda et al. 2002).

A different type of programme initiated in the USA has established an intellectual property clearing house to make information on the intellectual property owned by public research institutes, including universities, available to researchers around the world (Toenniessen 2000).

There are also suggestions that redesigning patent laws to narrow the type and scope of patent coverage ought to make more technologies accessible to public institutions. The thinking behind these suggestions is that applying a stronger standard for rejecting patent applications for inventions that are 'obvious' should deter the patenting of minor inventions. In addition, a law that requires an invention to be genuinely useful in theory should reduce the number of patent applications being submitted. At present, it is possible in some countries to submit patent applications for abstract concepts that potentially protect large areas of research and thereby exclude innovation by others.

Another option that may be attractive to developing countries is the creation of collaborations that involve research institutes, universities and the private sector (Khush 2003; Pray and Naseem 2003). The nature of these collaborations is likely to be influenced by the level of expertise and resources within the national public research institutes (Toenniessen 2000). Where a solid knowledge base exists, the public partners may be in a position to develop or acquire a technology that could be transferred into locally adapted varieties. Smaller institutes are more likely to provide the genetic resources and a positive public image. It is believed that such alliances would benefit public institutions and private companies; offering them an opportunity to license and distribute the technology (Barry and Horsch 1999).

The most well-known public–private sector partnerships include organizations such as the International Service for the Acquisition of Agri-biotech Applications (ISAAA) that negotiate access to private-sector technologies for the improvement of subsistence crops and/or the transfer of technology and know-how.

Although several types of public–private sector alliances already exist (James 1999;Salazar et al. 2000), the two newly established initiatives worth mentioning are the Global Cassava Partnership and the African Agricultural Technology Foundation (AATF), launched in November 2002 by FAO and in March 2003 by the Rockefeller Foundation, respectively. The former is a partnership involving some of the world's leading experts in cassava research, working mainly in public institutions. The AATF intends to function as a clearing house of available technologies with the primary aim of improving food security and reducing the poverty situation of small farmers, by facilitating transfer and use of appropriate technologies (Pray and Naseem 2003).

Such licensing arrangements have been put to the test in other fields (AATF 2005). However, as in the case of the AATF, a clearing house is required to acquire the necessary technologies and permit their further use for developing-country needs. The drawback may be a requirement to divide the commercial sector into subsistence, middle-income and commercial markets. This market division may not be easy to achieve as some large developing countries have both commercially important markets and subsistence farmers.

A briefing paper commissioned by the United Nations Industrial Development Organization (UNIDO) (Salazar et al. 2000) proposes six activities that build upon private-sector investment and enable biotechnology transfer:

1. enabling government policies;
2. access to up-to-date authoritative information;
3. regional brokering service to strengthen public–private partnerships;
4. a regional biotechnology investment service;
5. an international intellectual property escrow service; and
6. initiatives for risk-shifting.

Each of the above proposals can be implemented as a stand-alone project or in combination, as best suits the national and/or regional situation.

Access to Genetic Resources

Historically, plant genetic resources were freely provided by developing countries to gene banks worldwide. The resources in question did not belong to a particular individual, and are often considered a common heritage of mankind. The application of modern biotechnology to genes that could be incorporated in genetic resources of importance in rural populations raises concerns in that the small-scale farmers may have originally supplied the genetic resources for improvement. Once privately owned, these resources may be unavailable to the people who have ensured their conservation for centuries. Equally important is the aspect of access by researchers to genetic resources for further development on terms that recognize the contributions made by farmers to the conservation and sustainable utilization of these resources.

At the international level, the importance of national ownership of such resources is duly recognized. The International Treaty on Plant Genetic Resources adopted at an FAO conference in 2001 (FAO 2001c) provides the legal framework for dealing with the resources on which food security and sustainable agriculture depend. The Treaty gives a directive on the conservation and sustainable use of plant genetic resources for food and agriculture, making provision for the fair and equitable sharing of the benefits arising out of their use, in harmony with the principles of the Convention on Biological Diversity (CBD 2005a), but introducing the concept of farmers' rights.

In the discussions on farmers' rights, the main issues of concern revolve around benefit-sharing and prior informed consent, and the protection of traditional knowledge from 'biopiracy'. This means access to genetic resources must be on mutually agreed terms to promote their use and emphasize their importance to development. A number of organizations are discussing the protection of traditional knowledge and folklore.

The Treaty establishes a multilateral system of facilitated access and benefit-sharing (MLS) for key crops, emphasizing the interdependency of countries in terms of plant genetic resources for food and agriculture. Developing countries rich in genetic resources are encouraged to place germplasm in the MLS. The users of the material will sign a material transfer agreement, incorporating the conditions for access and benefit-sharing through a fund established under the Treaty. In return, the owners of the genetic resources will get a share of the benefits arising from their use and development in the way of information, technology transfer and capacity building. *Ex situ* material, collected before the entry into force of the CBD, does not come under the purview of the Convention, and would thus be dealt with under the Treaty. So far, 35 food and 29 feed crops have been entered into the system.

In principle, the genetic resources stored under this system are available for improvement to all interested researchers. Such wide-scale availability of germplasm has a potentially positive influence on access to improved technologies and nutritionally enhanced staple crops for the food-insecure. Genetic resources obtained from the MLS cannot be patented, even though it is not clear whether a gene isolated from such material may be protected or not.

Globalization

Globalization is a complex set of developments including trade liberalization, opening up of economies and integration of international markets (Diaz-Bonilla and Thomas 2001). It involves policy reforms in trade and reducing barriers to the international flow of goods, capital, labour, technology and ideas (UNDP 2003). Technological advancement is the driving force of improvements in world food production and the integration of the world economy. Knowledge in agricultural science depends on building onto experiences already gained. In this respect, the introduction of biotechnology ought to be perceived as nothing but a phase in the modernization of agriculture which started centuries ago. In its evolution, a global trend is emerging where fewer farmers produce more food. The global commercialization of agriculture has increased competition in domestic and international markets. Globalization has accelerated substantially since the 1980s, as the economies of developing countries have become subject to international market forces.

New technologies appear to drive the transformation of world food systems towards industrialized food processing, long-distance marketing and retail-business dominance. The benefits of globalization, however, seem to have bypassed low-income countries and, in particular, rural regions.

Even though the profit potentials were different, the merging of agrochemical and pharmaceutical divisions to form life sciences companies in the mid-1980s was a strategic synergy of R&D that allowed the production of new drugs, pesticides, GM crops and genetic treatments for disease (Chataway et al. 2002). This has led to new patterns of alliances with companies that develop routes to create and capture value.

The integration of economies cuts across all sectors and likewise is central to the emergence of new patterns of R&D. The phenomenon of globalization has raised concerns about the global acquisition of seed companies and mergers with chemical interests that have strategically strengthened their ability to market new products, and placed a substantial number of agricultural patents in the control of five major companies, namely Bayer, Dow Chemicals, Du Pont, Monsanto and Syngenta (Ching 2001; Graff and Newcomb 2003). The top 10 seed companies control one-third of the global seed market (RAFI 2000). This kind of consolidation has also placed much of the know-how under the control of these multinationals. To overcome problems of overlapping patent rights and reciprocal infringement lawsuits, companies make strategic alliances, mergers or acquisitions to obtain technologies in specified fields of research. Similar consolidation trends have been observed in developing countries, where PVP legislation is the only means of protecting improved seed — resulting in a proportionately small number of companies dominating global markets related to food and agriculture.

Overall, the seed industry can be divided into three segments: commercial seed, farm-saved seed, and public-supplied seed (James 1997). Farmers' sources of seeds tend to be flexible in that they respond to local needs and circumstances and therefore can greatly vary for the same crop according to location (Musa 1998). Farmers in most developing countries depend on farm-saved and public-supplied seeds. The latter provide seeds for the most important crops, while farm-saved seeds account for over 90% of the planted crops. Global consolidation of seed companies is likely to homogenize seed quality and limit choice (McGuire 1997). By the same token, the same genetic traits can be introduced into locally adapted varieties in different world regions. It is important, then, that when considering the effects of consolidation in the commercial seed market, it should be kept in mind that the commercial seed market is just one segment of the total seed market and, since it tends to concentrate on high-value seeds, provides less than 20% of the planted seed material in both developed and developing countries (Cromwell 1996). Furthermore, it should be recognized that GM seed represents only a minority of the commercial seed market, i.e. approximately 13% (James 2002b).

Market Access

More than 50% of the workforce in developing countries makes a living from agriculture, and so it follows that economic development in these countries relies on agricultural performance. The most important indicators of agricultural performance are an increase in domestic production and accelerated access to markets. Trade liberalization, as negotiated in the Uruguay Round, was intended to create opportunities that improve market access for the commodities of developing countries by eliminating agricultural subsidies, high import duties and other trade-distorting measures adopted by developed countries. At present, agriculture is subsidized at US$1 billion per day in the OECD countries, making it impossible for poor countries to compete on an equitable basis.

Limited market access represents the greatest hurdle to international trade, and consequently to technology access and acceptance (Juma and Konde 2002). If developing countries are to be integrated into the global economy, they must have something to offer that in turn translates into beneficial implications for food security. Policy options for those developing countries whose exports largely rely on agricultural trade will be influenced by the regulatory climate and consumer preferences in trading-partner countries, including agricultural-subsidy schemes that make the products of developing countries less competitive. The continued support by developed countries of high agricultural subsidies creates inequitable trading conditions for developing countries. In addition to the two major barriers to trade (tariffs and standards) for developing countries, a third hurdle in relation to the products of modern biotechnology may be the implementation of the CPB (CBD 2000). The CPB adopted in 2000, entered into force in September 2003. A legally binding instrument to its parties, it aims to regulate trade in products of modern biotechnology by protecting biodiversity and taking also into account effects on human health.

The backbone of the CPB is an advanced informed agreement that requires the party of export to seek consent from the party of import before the first shipment of living modified organisms intended for release into the environment. A simplified procedure is mandated for commodity trading. This is based on the proactive information exchange between the party of export and its potential trading partners via an Internet-based Biosafety Clearing House (CBD 2005c). The process is intended to facilitate trade by providing easy access to data and thereby enabling early assessments in prospective recipient countries. This, in effect, makes available information on the regulatory status of a new trait in any country before the initiation of trade or the delivery of food aid. Timely handling of this information by developing countries will be put to the test now that the CPB has entered into force. The SPS Agreement of WTO (WTO 1995) ensures that internationally traded food meets minimum standards based on the scientific principles set by the Codex Alimentarius Commission.

It is expected that when dealing with commodities containing products of modern biotechnology, the SPS Agreement will make reference to the Codex *Principles for the risk analysis of foods derived from modern biotechnology* (CAC 2003b) and the guidelines for risk/safety assessment of such foods (CAC 2003c,d). The Codex Alimentarius Commission approved the principles and guidelines in 2003.

The SPS Agreement both offers opportunities for improving food safety in developing countries and poses potential problems (Unnevehr 2001). The science and research-based requirements of the SPS Agreement can be substantial. Yet for countries that can implement the Codex standards for their export products, the opportunity to expand trade and acquire much-needed foreign currency is guaranteed. Those developing countries that cannot find resources to improve their food safety systems may find themselves excluded from international trade. More importantly, the limited capacity to do scientific risk assessment can compromise their ability to assess whether or not to import the products of modern biotechnology.

The increased international trade in food has created greater dietary diversity, year-round availability and often low prices in many countries. Non-traditional foods are becoming increasingly available worldwide. Producers in developing countries can benefit from expanded food exports which earn foreign exchange and increase rural incomes. These benefits, however, may not be realized if food safety and quality standards required in high-income markets cannot be met.

Capturing these new opportunities places the onus on developing countries to manage safety from farm to table (Wilson 2001). Testing of hazards at several points in the production process is expensive, and therefore prevention and control through documented production practices is often the only way to verify food safety. The concept of process control and hazard prevention is generally well understood in developed countries.

When a product is consumed domestically, and investments to meet export market standards for that product affect a large portion of production, those investments will have positive spill-over for domestic consumers. However, some products may be produced almost entirely for export, in which case investment to meet high standards of safety and quality in export will have little or no direct spill-over for domestic safety.

Nielsen and Anderson (2000) looked into the policy ramifications resulting from the economic effects of adopting GMOs. The study highlights the implications for the global economy when selected regions adopt modern biotechnology, in terms of agricultural production, trade and economic welfare. Analysis is made of the trade in cereal grains (excluding rice and wheat) and oilseeds. The authors suggest that similar analyses of crops unlikely to cause food-safety concerns in other areas and that are of potential economic importance to developing countries should be conducted. They believe prejudices in the more affluent countries may hinder the adoption and production of all products of modern biotechnology in all regions, unless more informed debates are held to discuss the potential opportunities for developing countries.

The United Kingdom Government's (2003) Strategy Unit also concluded (245) that trade, policies and consumer behaviour in the EU and United Kingdom would influence the decisions of developing countries related to the use of GM crops.

While export markets may be important for the rural poor, meeting higher standards may require additional management, capital investment, purchased inputs, monitoring and certification. Government focus on public-health issues will not necessarily address the export barriers.

Technologies in agricultural R&D have always been considered a major contributor to improving food security. Nonetheless, their successful application will depend on their relevance to poor people, the resolution of IPR disputes, and national and international, regulatory, trade, political and economic frameworks. The application of modern biotechnology in food and agriculture has the potential to reduce some problems associated with food insecurity. Many developing countries will need to overcome a number of obstacles before they can take full advantage of what modern biotechnology has to offer.

The development of products of modern biotechnology is capital-intensive as proprietary research tools must be licensed from the private sector in many systems. This situation has caused restrictions on innovations and is a barrier to the availability of the tools of research in both developed and developing countries. If developing countries are to rely on importation of new varieties, especially those developed with biotechnology, then allowing more flexible IPR standards makes good economic sense.

Research agendas in developing countries should focus on widening the crop base, and on enhancing the yield and nutritional value of the crops that are important to rural communities. Taking stock of national capabilities and prioritizing research objectives in line with the goals of sustainable agriculture will help put realistic opportunities into perspective for both conventional breeding and modern biotechnology techniques. A needs-driven technology is a tool for growth and development which the private sector is unlikely to undertake, because such crops are of low commercial value. Governments should take the responsibility of investing in public research that is crucial to reducing food gaps between rich and poor.

Depending on their goals and objectives, developing countries have choices with regard to the application of modern biotechnology:

(1)   leave its development to the private sector;

(2)   strengthen national public R&D capacity; and/or

(3)   create an enabling environment to build upon private-sector investment in public–private alliances.

The continued marginalization of developing countries from international trade will have a negative impact on the adoption and application of emerging technologies, including modern biotechnology. Therefore, a thorough consideration of all the issues pertinent to the application of a specific technology is critical to informed decision-making for the governments of developing countries. The use of new technologies in food and agriculture has become so politicized that regulatory institutions are obliged to provide assurance that deployment of such technologies will lead to improved nutrition and food security. Such policy and legislation cannot be developed in isolation and independently of international obligations and public opinion.

4

Use of Animals in Scientific Research

Animals for experiments should be procured by scientists from recognized animal facilities. The animals trapped from the wild, e.g. the monkeys, feral dogs and cats are also used in research as they are readily available and less expensive compared to colony bred animals.

These wild and feral animals are generally quarantined and stabilized in animal facility before use in experiments. The health and genetic status of these animals are not known and therefore a careful screening during quarantine is necessary. The wild and feral animals should be acquired after due clearance from Institutional Animal Ethics Committees and through certified suppliers.

The only authentic source of getting right type of animals for research should be from recognized scientific animal facilities where the animal colonies of known genetic and health status are available. Such animals only can provide reliable results.

The scientists should therefore insist upon getting defined animals through organized colonies eliminating unscrupulous traders, which not only supply poor stock of animals but also maintain these animals under most unethical and unhygienic conditions. A list of Scientific Institutions which maintain recognized animal strains. A directory of animal species and strains available with each of these institutions should be prepared and circulated. A list of 'Physiological Norms of Commonly Used Laboratory Animals' and 'Reproductive Data of Commonly Used Laboratory Animals' is given in Annexure 2 and 3 respectively.

Laboratory Animals

Systematic inbreeding and maintenance of inbred strains of laboratory mice and rats is of great importance for biomedical research. The inbreeding guarantees a consistent and uniform animal model for experimental purposes and enables genetic studies in congenic and knock-out animals. The use of inbred strains is also important for genetic studies in animal models, for example to distinguish genetic from environmental effects.

Evolution

Inbreeding has a variety of consequences. Allele exposure can cause genes to be expressed that are not otherwise expressed. This fact, combined with the fact that most mutations are recessive may indicate that inbreeding drives evolution. Speciation, a key process in evolution, depends on reproductive barriers, a necessary feature of which is inbreeding.

Line breeding

An intensive form of line breeding where an individual with highly desirable traits (S) is mated to his daughter (D1) and daughter's daughter (D2) and so on, in order to maximise the percentage of S's genes in the offsprings. The D3 offspring would have 87.5% of his genes while D4 offspring would have 93.75%. Such breeding methods can be used to create a "near clone" of a desirable individual. Line breeding is a form of inbreeding practiced by some animal breeders to "fix" desirable traits in a breed of animal, without as high a risk of producing undesirable traits that may occur with close inbreeding.

A typical example of line breeding would be what in human parlance would be considered a mating of first cousins or more distantly related individuals who share a common ancestor.

While line breeding is less likely to cause problems in the first generation than does inbreeding, over time, line breeding can reduce the genetic diversity of a population and cause problems related to a too-small genepool that may include an increased prevalence of genetic disorders and inbreeding depression.

Out crossing

Out crossing is the practice of introducing unrelated genetic material into a breeding line. It increases genetic diversity, thus reducing the probability of all individuals being subject to disease or reducing genetic abnormalities (only within the first generation). It actually can serve to increase the number of individuals who carry a disease recessively.

It is used in line-breeding to restore vigour or size and fertility to a breeding line. "Line-breeding", is where animals carry a common ancestor in their pedigrees and are bred together, should be considered distinct from the term "in-breeding" which is the production of offspring by parents more closely related than the average.

Out crossing is now the norm of most purposeful breeding, contrary to what is commonly believed. The out crossing breeder intends to remove the traits by using "new blood". With dominant traits, one can still see the expression of the traits and can remove those traits whether one outcrosses, line breeds or inbreds. With recessives, out crossing allows for the recessive traits to migrate across a population. It may actually increase the number of individuals carrying a disease. The out crossing breeder then may have individuals that have many deleterious genes that are expressed by placing their animals against a similarly outcrossed individual. There is now a gamut of deleterious genes within each individual in many breeds. However one may increase the variance of genes within the gene pool by out crossing, protecting against extinction by a single stressor from the environment. In cats, there is currently a study running to determine the genetic diversity within the cat breeds.

Out crossing is believed to be the "norm" in the wild. However, it is not logical as migration occurs by necessity. Feral cats, for example are one of the most inbred as individuals remain nearby their original homes, unless environmental stresses drive them to migration.

Breeders inbreed within their genetic pool, attempting to maintain desirable traits and to cull those traits that are undesirable. When undesirable traits begin to appear, breedings are selected to determine if a trait is recessive or dominant. Removal is accomplished by breeding two individuals of known genetic status, usually they are related.

In nature, where breeding is not managed, out crossing rates may be estimated by genetic analysis, by employing mathematical models of mating systems such as the mixed mating model or the effective selfing model. This allows calculation of the amount of genetic exchange between populations, and thus provides insights into the biogeography and phytogeography of species.

Gregor Mendel used out crossing in his experiments with flowers for his breeding stock. He then used the resulting offspring to chart inheritance patterns, using the crossing of siblings, and backcrossing to parents to determine how inheritance functioned.

Farm Animal Breeding and Reproduction in Europe – Characteristics:

- Farm animal breeding includes all animal species bred for a wide range of purposes.
- For breeding, animals meeting defined criteria are selected from animal populations.
- Farm animal species have been selected for desirable traits since they were first domesticated.
- Efficient reproductive techniques, such as artificial insemination, allow genetic improvement to be rapidly disseminated from the top of the breeding pyramid to benefit all producers and society as a whole.
- Animal breeding and reproduction are most effective when incorporated into herd and population management strategies.
- Balanced breeding requires a long-term sustainable vision developed jointly by breeders, scientists, and society.
- The added value of investments in genetics is cumulative.
- Breeding is knowledge intensive.
- An international farmers' organisation has established guidelines for data collection, farm management, and genetic evaluation to assist farmers and farmers' organisations.
- Breeding is society sensitive because it drives changes in the genetic makeup of animals and the use of new technologies (e.g. genomics, computing sciences).
- Genomics opens innovative prospects for sustainable animal production.
- A Code of Good Practice for Farm Animal Breeding and Reproduction Organisations is in place to encourage transparency and a dialogue of breeders with society.
- Breeding supports the health, feed efficiency, and welfare of farm animals as well as adequate management of animals and the environment.

The domestication of livestock species some ten thousand years ago was a vital step in the development of human civilisation. Over the centuries, domestication evolved into breeding and the genetic improvement of livestock. Nowadays, breeders measure many different animal traits and choose the best animals to be the parents of the next generation.

This leads to improvement generation after generation through the increased frequency of desired gene variants in the population. Because the breeding response is cumulative, permanent, and can be spread throughout the production chain, farm animal selection has a great impact on farm animal production. Europe has always played an important role in improving the world's major livestock and aquaculture species. European breeds are used across the world, and European farmers and breeding organisations are major players on the global market. The European farm animal breeding sector will thus have a great influence on, and therefore responsibility for, the future genetic makeup and characteristics of farm animal populations worldwide.

A conservative estimate of the economic gain achieved each year by animal breeding at farm level is €1.83 billion in Europe alone. Hence, the genetic gain achieved by breeders is carried over to producers as an economic gain reaching approximately 1.5% of the economic value of EU farm animal production.

Laboratory Animal Husbandry and Management

Housing and Environment

Laboratory animals are very sensitive to their living conditions. It is important that they are housed in an isolated building located as far away from human habitations as possible and not exposed to dust, smoke, noise, wild rodents, insects and birds. The building, cages and environment of animal rooms are the major factors which affect the quality of animals. In planning an animal facility the space should be well divided for various activities. The animal rooms should occupy about 50-60% of the total constructed area and the remaining area should be utilized for services such as stores (8-10%), washing (8-10%), office and staff (8-10%), machine rooms (4-5%) quarantine and corridors (12-15%).

The cages should be made of suitable metal (stainless steel, galvanized iron sheet/rods) or synthetic material (polypropylene/

polycarbonate). They should be of suitable size for each species of animal and should have adequate arrangement for feeding and watering. They must be free from crevices, corners and sharp edges for easy cleaning and to avoid injury. The bedding should be of right material and sterilized before use. Common bedding materials used in India are paddy husk, saw dust, paper currings, dry grass and crushed corn cobs.

The environment of animal room (Macro-environment) and animal cage (Micro-environment) is an important factor on which the production and experimental efficiency of the animal depends. Since animals are very sensitive to environmental changes, sharp fluctuations in temperature, humidity, light, sound and ventilation should be avoided. The recommended environmental requirements for animal rooms.

A constant room temperature is essential, because variation in room temperature causes change in food and water intake. A charge in temperature of 4°C can cause 10-fold alteration in biological responses. The temperature also affects fertility and lactation. Coupled with high humidity the increase in temperature causes ammonia built up. If the ventilation is not proper the high ammonia concentration causes respiratory irritation to both animals and attendants, predisposing them to infection by lowering their resistance. An effective ventilation system with 10-12 air changes per hour of 100% fresh air must be provided for animal rooms.

Light and sound are other important factors. The light intensity, the wave length and the photo cycle affect the health and behaviour of the animals. Sudden and sharp sounds in the animal rooms disturb the health and behaviour of animals and may give rise to ear damage, hypertension, cannibalism, etc.

Breeding and Genetics

For initiating a colony, the breeding stock must be procured from a reliable/accredited source, ensuring that genetic make-up and health status of animals is known. In case of an inbred strain, the characters of the strain with their gene distribution and the number of inbred generation must be known for further propagation. The health status should indicate their origin, e.g. conventional, specific pathogen free or transgenic gnotobiotic or knock-out stock. The known nutritional status and feeding habits of the stock are also of advantage.

The animal colonies may be inbred or outbred. In case of inbred colonies the number of generation of brother x sister mating and latest genetic monitoring parameters for various markers should be known. Mutations or genetic contamination can be detected by using screening methods, such as histocompatability (skin grafting), biochemical markers, coat colour studies, mandibular biometry or immunological studies. Phenotypic visual characters may also sometime provide clue of genetic contamination.

Transgenic/Knock-out Animals

Transgenic animals are those animals into whose germ line foreign gene(s) have been engineered, whereas knock-out animals are those whose specific gene(s) have been disrupted leading to loss of function. These animals can be bred to establish transgenic animal strains. Transgenic animals are used to study the biological functions of specific genes, to develop animal models for diseases of humans or animals, to produce therapeutic products, vaccines and for biological screening, etc. These can be either developed in the laboratory or procured fpr R&D purpose from scientific/academic institution or commercial firms, generally from abroad. Those laboratories developing transgenic animals should pay special attention to the following points :-

1. Photoperiod is a critical regulator of reproductive behaviour of many species of animals. Inadvertent light exposure during the dark cycle should be minimized or avoided. A time controlled lighting system should be used to ensure regular diurnal cycle.

2. Embryo transfer has to be carried out using anesthetics.

3. Pseudopregnant females which receive embryos should be kept in separate rooms where there is no disturbance.

4. Bedding, feed, water or cage should not be changed for about 3-4 days after embryo transfer as at this stage there is high risk of embryo resorption and termination of pregnancy.

5. The bedding changes and handling of the female should be carried out by skilled caretaker till delivery and weaning is over.

6. A high protein diet should be given to the lactating mother.

Maintenance : Housing, feeding, ventilation, lighting, sanitation and routine management practices for such animals are similar to those for the other animals of the species as given in the guidelines. However, special care has to be taken with transgenic/gene knock-out animals where the animals can become susceptible to diseases or where special conditions of maintenance are required due to the altered metabolic activities. The transgenic and knock out animals carry additional genes or lack genes compared to the wild population of the species, and therefore to avoid the spread of the genes in wild population, neither they should inadvertently cross breed with other animals or be released in the wild. Special care should be taken to maintain those animals which have been genetically manipulated to produce models for diseases of humans or other animals. The transgenic and knockout animals should be maintained in clean room environment or in animal isolators.

Disposal : The transgenic and knock-out animals should be first euthanasised and then disposed off as prescribed elsewhere in the guidelines. A record of disposal and the manner of disposal should be kept as a matter of routine.

The transgenic and knock-out animals need greater level of monitoring than other animals as transgenes might have unexpected effect on the phenotype and its interaction with the environment.

Nutrition and Feeding

The results of an experiment are likely to be influenced by co-existence of nutritional deficiencies and imbalance. It is, therefore, essential that laboratory animals are maintained on a balanced diet based on nutritional requirements of each species. Special care is needed on nutritional elements, ingredients used in diet, and feeding practices. A balanced diet should contain protein, carbohydrates, fat, minerals, vitamins, roughage and water in required proportions for each species of animal. These requirements for commonly used species. Only quality ingredients should be used in a diet and they should be free from dust, moulds, fungi and other contaminants. Each animal must get required quantity of feed, based on animal maintenance and production requirements. The feed should be palatable so that it is consumed in adequate quantity by the animals. Any undesirable odor always causes under consumption resulting in nutritional deficiency in the animals. No drug, hormone or antibiotic should be added in the feed as these are likely to disturb the normal metabolism of the animals and produce biased results.

The ingredients and the prepared feed must be stored and handled carefully so as to avoid any contamination. The food must be of right consistency and should be presented to animals in proper type of hoppers to avoid wastage. In some cases the feed may be divided in 2-3 meals during the day. Pelleted feeds balanced for different species of animals are now available commercially. These are easy to procure and use without wastage. However one has to be careful on quality of the feed from batch to batch. It should be obligatory on manufacturer to mark each bag with the type of food, date of manufacture, the batch number, the ingredients used and chemical composition. Random chemical analysis must be carried to for major nutrients to monitor the quality of food from time to time. Clean, chlorinated water should be available to the animals ad lib.

Hygience and Disease Control

The building for housing the animals should be provided with barriers to control the entry of contamination into the building through men, material and wild animals. Strict barriers should be provided to avoid the entry of wild rodents, birds, insects and pests. Visitors and service staff should be allowed entry with care and when necessary. On the exit side an efficient monitoring service should be established to monitor the prevalence of any infection in the colony. A regular medical check up of the staff, postmortem of dead and sacrificed animals and screening of waste material of the rooms are essential.

Personnel and Training

The selection of animal facility staff, particularly the staff working in animal rooms or involved in transportation, is a critical component in the management of an animal facility. The staff must be provided with all required protective clothing (masks, aprons, gloves, gumboots, etc.) while working in animal rooms. Facilities should be provided for change over with lockers, wash basins, toilets and bathrooms to maintain personal hygiene. It is also important that a regular medical check-up is arranged for the workers to ensure that they have not picked up any zoonotic infection and also that they are not acting as a source of transmission of infection to the animals.

He should ensure that persons working in animal house don't eat, drink, smoke in animal room and have all required vaccination, particularly against tetanus and other zoonoses. Initial in-house training of staff at all levels is essential. A few weeks must be spent on the training of the newly recruited staff, teaching them the animal handling techniques, cleaning of cages and importance of hygiene, disinfection and sterilization. They should also be made familiar with the activities of normal healthy and sick animals so that they are able to spot the sick animal during their daily routine check up of the cages. At national level suitable training programmes should be organized by the National Centres to provide training in care, breeding, management, handling of animals for the staff working in animal breeding and holding units.

Orientation training programmes should also be initiated for the investigators working in different areas to acquaint themselves with various experimental techniques. Such a course should address the undermentioned topics :

a) biology and husbandry of laboratory animals

b) genetic make-up

c) microbiology and diseases

d) health-hazards in the animal house

e) anesthesia, analysis and experimental procedures

f) alternatives to animal use

g) ethical aspects and legislation.

The training courses and workshops may also be organized for the senior level biological scientists to evoke awareness among them about the use of animals in research, alternatives available and the ethical and legal provisions in regard to use of animals. This is particularly important care and management for the supervisory staff and veterinarians exists in the country. The national level training programmes of following type are essentially required.

Good quality animals are those which are free from disease. Animal of a specified strain should also have all the characteristics of that strain, i.e. they should be genetically anthesised. The results of regular monitoring of parameters of genetic purity must be scrupulously recorded. Proper record-keeping is extremely important and vital for an animal facility. The forms should be simple but complete and preferably computer compatible. Too exhaustive and unnecessary recording should be avoided as these are not useful. Records of breeding and experimentation and deaths of all experimental animals at various stages are essential. Receipt and issue of food and other stores should be recorded. Log books of various machines such as incinerator, boilers, air-conditioning plant should be maintained. Monthly and annual reports of the activities should be prepared and reviewed for evaluation of work and future planning.

Experimentation and Veterinary Care

The experimental animal units should generally be looked after by qualified investigators. These units must have adequate housing and technical facilities for experiment and post-operative care. The equipment provided in the experimental unit should be appropriate for the needs of the experiments. No technique should be used which may cause avoidable discomfort to the animals. The post-operative holding rooms and cages should be comfortable and such animals should remain under the care and supervision of an experienced scientist or a qualified veterinarian. The person actually incharge of animal facility should preferably be a veterinarian or a person qualified in laboratory animal science. In any case an experienced veterinarian must be readily available in an animal holding for health care, monitoring, diagnosis and treatment of diseases and injuries. A veterinarian could also be helpful to investigators in animal anaesthesia and surgery.

Transport of Laboratory Animals

The transport of animals from one place to another is very important and must be undertaken with care. The main considerations for transport of animals are, the mode of transport, the containers, the animal density in cages, food and water during transit, protection from transit infections, injuries and stress. The mode of transport of animals depends on the distance, seasonal and climatic conditions and the species of animals. Animals can be transported by road, rail or air taking into consideration above factors. In any case the transport stress should be avoided and the containers should be of an appropriate size so as to enable these animals to have a comfortable, free movement and protection from possible injuries. The food and water should be provided in suitable containers or in suitable form so as to ensure that they get adequate food and more particularly water during transit. The transport containers (cages or crates) should be of appropriate size and only a permissible number of animals should only be accommodated in each container to avoid overcrowding and infighting.

Anaesthesia and Euthanasia

The scientists should ensure that the procedures which are considered painful are conducted under appropriate anaesthesia as recommended for each species of animals. It must also be ensured that the anaesthesia is given for the full duration of experiment and at no stage the animal is conscious to perceive pain during the experiment.

If at any stage during the experiment the investigator feels that he has to abandon the experiment or he has inflicted irreparable injury, the animal should be sacrificed by overdose of anaesthetic. Neuromuscular blocking agents must not be used without adequate general anaesthesia. In the event of a decision to sacrifice an animal on termination of an experiment or otherwise an approved method of euthanasia should be adopted and the investigator must ensure that the animal is clinically dead before it is sent for disposal.

Anaesthes

Unless contrary to the achievement of the results of study sedatives, analgesics and anaesthetics should be used to control pain or distress under experiment. Anaesthetic agents generally affect cardiovascular, respiratory and thermo-regulatory mechanism in addition to central nervous system. Before using actual anaesthetic the animal is prepared for anaesthesia by over night fasting and using pre-anaesthetics, which block parasympathetic stimulation of cardio-pulmonary system and reduce salivary secretion. Atropine is most commonly used anti-cholinergic agent. Local or general anaesthesia may be used, depending on the type of surgical procedure.

Local Anaesthesia : Local anaesthetics are used to block the nerve supply to a limited area and are used only for minor and rapid procedures. This should be carried out under expert supervision for regional infiltration of a surgical site, nerve blocks and for epidural and spinal anaesthesia.

General Anaesthesia : A number of agents are used as inhalants. General anaesthetics are also used in the form of intravenous or intra-muscular injections such as barbiturates. Species characteristics and variation must be kept in mind where using an anaesthetic. Side-effects such as excessive salivation, convulsions, excitement and disorientation should be suitably prevented and controlled. The animal should remain under veterinary care till it completely recovers from anaesthesia and post operative stress.

Euthanasia

Euthanasia means "easy death" and is resorted to in events where an animal is required to be sacrificed on termination of an experiment or otherwise for ethical reasons. The procedure should be carried out quickly and painlessly is an atmosphere free from fear or anxiety. For accepting of an euthanasia method as humane it should have an initial depressive action on the central nervoussystem for immediate insensitivity to pain. The choice of a method will depend on the nature of study, the species of animal and number of animals to be sacrificed. The method should in all cases meet the following requirements :-

(a) Death, without causing anxiety, pain or distress with minimum time lag phase.

(b) Minimum physiological and psychological disturbances.

(c) Compatibility with the purpose of study and minimum emotional effect on the observer and operator.

(d) Location should be separate from animal rooms and free from environmental contaminations.

(e) Method should be reliable, reproducible and safe to the personnel involved.

(f) Simple and economical.

It is recommended that tranquilizers be administered to larger species such as monkeys, dogs and cats before an euthanasia procedure. A number of euthanasia methods have been recognized humane which could be physical, use of inhalant gases, injectable drugs and general anaesthetics in heavy dose.

Disposal of Animal Carcasses

All animal carcasses whether healthy, infectious or radioactive, must be packed in polythene bags before sending them for disposal. All healthy or infectious animals may be buried deep in the ground covered with lime and disinfectants or burnt in an incinerator. Animals with radioactive material should be packed in double polythene bags and dumped in a special pit meant for this purpose. Details of the pit can be obtained from Bhabha Atomic Research Centre, Mumbai. Strict precautions should be taken to safeguard the health of the personnel handling infectious and radioactive material and in no case these should be brought out in open containers for disposal.

The investigators working with infectious and radioactive material must ensure that the animals are properly disposed off on termination of their experiment and that infection is not transmitted to other animals in the room or to the personnel involved in handling such animals during the experiment and also at the time of disposal. The staff handling such animals for disposal must be apprised of the hazards involved in their job and protective clothing, gloves and mask must be provided to them foro their personal safety.

Laboratory Animal Ethics

All scientists working with laboratory animals must have a deep ethical consideration for the animals they are dealing with. From the ethical point of view it is important that such considerations are taken care at the individual level, at institutional level and finally at the national level. Individually each investigator has an obligation to abide by all the ethical guidelines laid down in this regard at institutional level. The Head of the Institution, maintaining animals for scientific experiments, should constitute an Animal Ethics Committee for experimentation to ensure that all experiments conducted on animals are rational, do not cause undue pain or suffering to the animals and only minimum number of animals are used. The constitution and terms of reference of the Animal Ethics Committee should be well defined.

An Animal Ethics Committee should include : a senior biological scientist of the Institute, two scientists from different biological disciplines, a veterinarian involved in care of animals, the scientist incharge of animal facility, a scientist from outside the institute, a non-scientific socially aware member and a member or nominee of appropriate regulatory authority of Govt. of India. A specialist may be co-opted while reviewing special projects using hazardous agents such as radioactive substance and deadly micro-organisms etc. The investigator may also be called in for any clarification, if required.

The Animal Ethics Committee has to examine all projects involving use of animals before implementation, to ensure that minimum number of animals is used in the project and the ethical guidelines are strictly adhered to. It will also examine that the scientists and technicians handling animals possess adequate skill to perform the experiment. All animals will be maintained under standard living conditions and experiments will be conducted with care. All invasive experiments will be conducted under proper anaesthesia and on termination of an experiment, the animal will be humanely sacrificed under anaesthesia. Before disposal it must be ensured that the animal is clinically dead.

Ethical Guidelines for Use of Animals in Scientific Research

1. Animal experiments should be undertaken only after due consideration of their relevance for human or animal health and the advancement of knowledge.

2. The animals selected for an experiment should be of an appropriate species and quality, and minimum number should

be used to obtain scientifically and statistically valid results.

3. Investigators and other personnel should treat animals with kindness and should take proper care by avoiding or minimizing discomfort, distress or pain.

4. Investigators should assume that all procedures which would cause pain in human beings may cause pain in other vertebrate species also (although more needs to be known about the perception of pain in animals).

5. Procedures that may cause more than momentary pain or distress should be performed with appropriate sedation, analgesia or anaesthesia in accordance with accepted veterinary practice. Surgical or other painful procedures should not be performed on unanaesthetized animals.

6. At the end of, or when appropriate during an experiment, the animal that would otherwise suffer severe or chronic pain, distress, discomfort, or disablement that cannot be relieved or repaired should be painlessly killed under anaesthesia.

7. The best possible living condition should be provided to animals used for research purpose. Normally the care of animals should be under the supervision of a veterinarian or a person having adequate experience in laboratory animal care.

8. It is the responsibility of the investigator to ensure that personnel conducting experiment on animals possess appropriate qualifications or experience for conducting the required procedures. Adequate opportunities have to be provided by the institution for inservice training for scientific and technical staff in this respect.

9. In-vitro systems to replace or reduce the number of animals should be used wherever possible.

In-Vitro Systems to Replace Animals

A number of in vitro systems can be used to reduce/replace animals in experimentation. These systems could be the living or the non-living systems. The living systems are tissue and organ culture, lower animals and microorganisms and human volunteers in restricted cases. The nonliving systems could also be used in place of animals in certain areas and these include chemicals, mechanical models, mathematical models, computer simulation, DNA recombinant technology and synthetic substances.

5

Statistical Methods in Animal Breeding

Introduction

A recent GIFT workshop had two papers that discussed future dairy cattle research. The two papers were in good agreement of the future statistical needs. These included methods for test day models, international comparisons, non additive variance, non-linear models and individual gene models. They also highlighted a trend to more sophisticated analysis leading to less biased predictions and more progress at the expense of greater variance or risk. Cost of analysis was suggested to be small compared to the cost of collection of data. There was a concern that uncertainties in parameters might erode possible gains. There was also a hope that prediction were robust to bad luck. Variance parameter estimation plays an integral role in several of these topics. We therefore intend to review this area hoping to identify themes that will lead to more rapid change. Frem Narain has made an outstanding contribution to statistical genetics, including the application of statistics to plant and animal breeding, so we think our small paper contribution is particularly apt to this volume in honour of Prem's 70th birthday

Using PLI to Breed

In the last 10 years, genetic indices for dairy cattle across the world have been moving toward the breeding of a more sustainable cow with a shift towards lifetime production rather than lactation production. Consequently, the UK, has moved from Profit Index (£PIN) which was totally production based to Profitable Lifetime Index (£PLI) which also includes non-productive traits such as fertility, lifespan and somatic cell count.

Genetic trends in dairy cows have been for increasing milk production but this has led to declining genetic merit for butterfat and protein per cent, fertility and lifespan. Infertility is estimated to cost the average 100-cow herd around £22,000 per year and is the primary reason for dairy cow culling.

However, when selecting AI sires, many farmers do not take account of sire fertility index, despite the fact that DairyCo has published it with sire proofs for over five years now.

This is the message from Marco Winters, head of genetics for DairyCo Breeding+ who highlights that: "There is a need for farmers to gain a better understanding of breeding and sire selection so that a clear breeding plan can be established."

"As farmers have used bulls with higher and higher Predicted Transmitting Abilities (PTAs) for milk production, their PTAs for calving interval have got worse. In other words, they've got longer," says Mr Winters.

"However, since 2005 when the Fertilisty Index was first launched, selection continued to favour milk production but the genetics for calving interval started to improve. This is excellent news, because it tells us that it's possible to continue breeding for better production without sacrificing fertility and offers the realistic hope that future improvements can be made through breeding."

To investigate the practical impact of sire fertility index on daughter fertility, Dr David Mackey of CAFRE has retrospectively analysed the fertility records of Greenmount's Future Herd and several other pedigree and non-pedigree herds throughout Northern Ireland.

"While it takes data from a lot of cows to demonstrate genetic trends, the benefits of selecting sires with a positive fertility index can be clearly seen through both a reduced calving interval and an increase in conception rate."

Dr Mackey goes on to comment that: "Had we know that some of the sires used before 2005 carried a negative Fertility Index, they would not have been selected. Since then only bulls with a positive Fertility Index are selected and we should start seeing the benefits of this in the next few years." A number of dairy farmers in Northern Ireland actively use £PLI in their sire selection each year and see the benefits of it, including Drew McConnell from Newtonstewart, Gary McHenry from Lurgan and Kenneth Montgomery from Eglinton.

Drew comments that he "Always selects bulls on £PLI, ideally those with a £PLI of £180 or more," but goes on to say that "they need to have positive milk figures and positive fertility as well."

Analysis of Drew's breeding records established that, on average, each one-point increase in a sires Fertility Index reduced the average calving interval of his daughters by almost a day, twice that estimated by DairyCo (0.5 days).

This demonstrates the benefit of including positive fertility in your sire selection criteria. Drew also says that "I cull a lot of cows for mastitis and I'm trying to address this through breeding by selecting bulls with a negative SCC score."

Similarily, Gary McHenry focuses on £PLI as a simple and effective screening tool.

"As we already have the genetics to produce 9,200 litres, our focus isn't so much on milk yield but on protein percentage, fertility and somatic cell count.

"We only look at bulls with a PLI of £150 or more, and then select bulls with protein percentage above +0.10 per cent, fertility index above five and SCC less than-10. These traits are all heritable and will contribute to higher profitability in future."

Mr Winters comments: "Farmers like Drew and Gary who are paying attention to and adopted the strategy of selecting on fertility are to be congratulated, and the AI companies have played an important role in the process through the increased listing of positive fertility bulls in their catalogues."

However, fertility isn't the only non-productive trait considered within the £PLI index. Bulls also carry proofs for Lifespan and Somatic Cell Count, and these should also be considered when selecting bulls.

Mr Winters suggests that "Sires should be screened on £PLI in the first instance, and then check the fitness traits important in your herd and ensure you are using bulls that have strengths in the right areas. For many herds, the area which will need particular improvement is fertility, and today, it's possible to find more bulls on the market than ever before which will transmit good fertility on to their daughters."

Advancing Genomic Selection

Genomic selection supplements the traditional method of genetic evaluation with information on the DNA of an animal to improve the accuracy of identifying genetically elite animals. Noirin McHugh and Donagh Berry from Teagasc say that greater gains can be achieved by applying genomic selection to females as well as males.

Uptake of Genomic Selection in Ireland

Since its initial launch in 2009 the uptake of genomic selection in Ireland has been strong. In 2009 genomically selected (GS) bulls accounted for 34 per cent of the total AI straws sold in Ireland; 2010 saw this figure increase further to 40 per cent. One of the main reasons for the rapid adoption of this technology by Irish dairy farmers is the greater EBI of the GS bulls (218 in 2010) compared to the Irish daughter proven bulls (€146 in 2010).

The reliability of available young bulls increased from 32 per cent prior to genomic selection to 54 per cent with genomic selection. Although the reliability of the GS bulls is lower than for most Irish proven bulls, the associated increase in the ebI of the Gs bulls has seen a strong uptake among Irish farmers. However, as the reliability of the GS bulls remains relatively low, farmers are adhering to the advice given and using an average of four Gs bulls per herd to spread their risk.

Advances In Genomic Selection in Ireland

Fundamental to generating a benefit from genomic selection, is knowledge on the DNA profile most suited to Ireland. A population of Irish proven animals, commonly known as a training population, is used to relate the DNA profile of the individual animals to their genetic merit for the array of traits evaluated by the ICBF.

Through international collaboration, the training population in Ireland has more than quadrupled from 945 domestically proven animals in 2009 to 4,196 in 2011. This training population size is still lower than in some countries and is reflected in lower increases in reliabilities from genomic selection. When genomic selection was initially launched in 2009 the service was only available for animals that were at least 50 per cent Holstein due to the small size of the training population for Friesians.

Genomic selection, however, is now available for Friesians. other recent advances include the development of genomic selection for linear type traits; initial results show an increase in the reliability of the type traits by approximately 10 percentage units.

Genomic Selection on Farm

Genomic selection is now available to all dairy farmers. ICBF provide two genotyping services: the low density "3k" (i.e. 2,900 markers) platform or the normal density "50k" (54,001 markers) panel. because animals inherit large chunks of their dna from their parents, animals that have both their sire and maternal grandsire genotypes in the ICBF database (now 9,000 genotypes) can avail of the lower-cost 3k genotype service; while other animals can avail of the 50k genotype service; the cost of each service is €50 and €99, respectively. Farmers can log onto the ICBF website and see which animals can avail of the lower cost service.

The Use of Females To Exploit Genomic Selection

To date the focus of genomic selection has been on the male side; however, the female has an increasingly important role to play. Studies have shown that the genotyping of females can substantially increase the size of the training population thereby improving the accuracy of the genomic ebIs. The genotyping of females can potentially result in a three fold increase in the rate of genetic gain. An economic analysis con irms that genotyping females can result in very high rates of return on the investment.

Although genomic selection has progressed rapidly since its inception, there are still many other areas where further gains can be made. Our continued role in international collaborations ensures that Ireland remains at the forefront of this cutting edge technology.

The swapping of genotypes with international partners ensures that the training population will continue to grow thereby increasing the reliability of our genomic EBIs. Access to more DNA markers may increase genetic gain further and expedite its application to multiple breeds.

The success of genomic selection has been due to the support and involvement of the entire industry, from research carried out by Teagasc, its implementation by the ICbF, the investment in infrastructure at Weatherbys to generate the DNA Profile, and the large uptake by dairy farmers. Teagasc and the ICBF will continue to work closely with the industry to further enhance genomic selection and ensure that the benefits are realised at farm level.

Peering at Genes to Detect Origin of Cattle Diseases

The eyes are said to be the windows into the soul of humans, but in the case of cattle, they may hold clues to overall animal health. A closer look at pinkeye is offering insight into other costly bacterial diseases as well.

Researchers at the Agricultural Research Service's Roman L. Hruska US Meat Animal Research Centre (USMARC) in Clay Centre, Nebraska, have found that genes are linked to the incidence of several diseases.

Eduardo Casas, a geneticist in the Genetics and Breeding Research Unit at the time of the study, and former ARS scientist Gary Snowder discovered a quantitative trait locus (QTL), or location, on bovine chromosome 20 that is associated with pinkeye, foot rot, and bovine respiratory disease.

Cow receiving treatment for pinkeye at the U.S. Meat Animal Research Centre, Clay Centre, Nebraska.

Looking Beyond the Pathogens

Scientists have known for some time what causes pinkeye, also known as "infectious bovine keratoconjunctivitis," and other diseases. The bacteria Moraxella bovis is the most common pathogen associated with pinkeye.

Pathogens associated with bovine respiratory disease include viruses, such as bovine viral diarrhea virus, bacteria, like Mannheimia hemolitica, and mycoplasma. Fusobacterium necrophorum and Porphyromonas spp. are the main bacterial pathogens for foot rot, or infectious pododermatitis.

"Scientists have spent a lot of effort and money studying the pathogens that make animals sick," says Mr Casas, who is now research leader for ARS's Ruminant Diseases and Immunology Research Unit in Ames, Iowa.

"We've made a lot of progress, but the microbes are still around. Therefore, the main focus of this research was to look at diseases from the animal's point of view."

Mr Casas' approach was to examine the genetic makeup of cattle for evidence of genes associated with conferring resistance or tolerance to diseases. His initial study focused only on pinkeye because it's easy to see and measure in cattle, he says.

Different breeds vary in their pinkeye tolerance. For example, Herefords are very susceptible, but Brahmans are highly resistant. With this in mind, a Brahman-Hereford crossbreed sire was mated to other breeds to yield more than 540 offspring.

USMARC, which has more than 6,000 head of cattle, provided an ideal location to study different breeds affected by pathogenic diseases. (K7648-3)

"This particular bull was heterozygous for all genes that would confer tolerance to pinkeye," Mr Casas says.

"Half of the offspring inherited the resistant gene, and the other half inherited the susceptible gene."

When scientists looked at 36 offspring affected by pinkeye, they found that regions on chromosomes one and 20 harbored genes that influence the presence of bacteria, but no strong linkage to a QTL was identified. So the team took a different approach.

A Tale of Three

Following up on a theory that the immune system is influenced by various genes, Mr Casas and Mr Snowder conducted a second study. They combined the incidences of three highly prevalent bacterial diseases affecting feedlot cattle—pinkeye, foot rot, and bovine respiratory disease.

"When you put all three diseases together, you're looking at the overall health of the animal, or resistance to multiple diseases, rather than a disease-specific response," Mr Snowder says.

"In other words, the particular loci affecting an individual disease may not be easy to pick up, but it might be easier to pick up markers that are related to the general health of the animal."

Selection for disease resistance is one of several possible interventions to prevent or reduce economic loss associated with animal disease and to improve animal welfare, according to Mr Casas and Mr Snowder.

A common condition affecting breeding-age beef heifers, pinkeye has a marked economic impact on the cattle industry—costing an estimated $150 million a year due to lower weight gains, decreased milk production, and treatment. Although not fatal, this highly contagious disease can affect up to 80 per cent of a herd. Calves being weaned are even more susceptible and can lose as much as 10 per cent of their body fat if they contract the disease. Bovine respiratory disease—pneumonia—is the most common and costly feedlot disease in the US. It accounts for 75 per cent of feedlot morbidity and up to 70 per cent of all deaths. Economic losses to cattle producers exceed $1 billion annually from animal deaths, reduced weight gain, lower feed efficiency, treatment costs, and poor-quality meat and hide products.

While foot rot is not as expensive as other diseases, it is estimated to cost dairy producers about $120 to $350 per animal. Foot rot causes lameness and leads to reduced milk yields, lower reproduction performance, increased involuntary cull rates, and discarded milk.

Producers have been managing these diseases with various treatments and management practices. USMARC, which has more than 6,000 head of cattle, provided an ideal location to study different breeds affected by pathogenic diseases.

Geneticist Eduardo Casas (right) and technician Sandra Nejezchleb prepare bovine DNA samples for mass-spectrometry. Results are used to assess whether changes in DNA sequence are associated with pinkeye, also known as "infectious bovine keratoconjunctivitis." (D2236-9)

Breeds Apart

In addition to the Brahman-Hereford family studied in the pinkeye experiment, three other half-sibling families were produced to detect QTLs associated with combined incidences of the three diseases.

The second half-sibling family was developed from a Brahman-Angus sire and produced 176 offspring. A Piedmontese-Angus sire fathered 209 calves, and a Belgian Blue-MARC III (part Red Poll, Pinzgauer, Hereford, and Angus) sire produced 246 offspring.

Researchers used microsatellite markers—short, repetitive DNA sequences used as genetic markers to track inheritance—to screen the genome of each family. Informative markers were chosen within a family based on their location in each chromosome.

All animals were observed daily throughout their life-span for pinkeye, pneumonia, and foot rot and treated when symptoms occurred. The 240 calves infected by one or more of the diseases were classified as affected by a microbial pathogenic disease and coded. Analysis of DNA blood samples taken from these animals revealed QTLs for disease activity.

Though scientists have discovered genetic locations that may influence resistance or susceptibility to bacterial diseases, there's more to do. "We don't know what the gene or genes are yet, and that's what we are working on," Mr Casas says. More study needs to be done to confirm the association between the genes and disease."

"What's interesting about the markers on chromosome 20 is that they are in very close proximity to other markers related to other diseases. That particular region may have a significant effect on the general health of animals," Mr Casas says.

Additional studies are under way to detect genes associated with reduced susceptibility to bacterial diseases, including Johne's disease and bovine viral diarrhea.

"The costs for treating animals that have these diseases are enormous," Mr Casas says.

"Identifying genes responsible would provide an opportunity for effective crossbreeding to produce animals with increased disease tolerance, which would greatly reduce the economic impact to the cattle industry."

Boosting Breed Plan

Three funding terms of Beef CRC have ensured that new, hard-to-measure but economically very important traits have been added to BREEDPLAN, significantly increasing BREEDPLAN's value to the beef industry. Prediction equations based on DNA information from the new high density panels will be delivered to BREEDPLAN

Since the Beef CRC began contributing data to BREEDPLAN in 1993, performance-recorded Australian cattle breeds have shown steadily rising rates of genetic gain. Now a wave of new genetic information is coming through, made possible by the inclusion of new DNA-based genomic technologies.

Breedplan is built on "phenotypic" data, which comprises measured records of economically important characteristics of an animal. About a decade ago, technologies began appearing that seemed to promise to greatly reduce the enormous labour and cost of phenotypic recording by identifying the genes associated with these economically important traits.

At the time, "genotypic" (DNA-based) data promised a revolution that would leave the old methods in the dust. However, according to one of BREEDPLAN's founders, Dr Hans Graser, Director of the Animal Genetics and Breeding Unit (AGBU), the process has proved not to be so simple. There is now a clear realisation that gene interactions are far more complicated than was originally thought.

Forecasts of dramatic advances through genomics were made on the assumption that a handful of genes would explain most of the genetic variation for a trait.

However once the Bovine Genome was sequenced in 2006 and the full assembly released in 2009, it became clear that things were not as straightforward as expected. It turns out,Dr Graser says, that hundreds, even thousands of interacting genes may be at work to control a single trait.

Technology has produced some answers to this complexity, including the single-nucleotide polymorphism (SNP, pronounced "snip") chip. These chips allow tens or hundreds of thousands of genetic sequence differences to be quickly compared against each other, revolutionising the business of gene discovery.

The Beef CRC has used this technology to record genetic differences from thousands of animals, even reaching back into its archival material to genotype long-dead cattle used in Beef CRC's first term. A few CRC animals have also been fully sequenced, a procedure that was prohibitively expensive only a few years ago, to contribute to the global knowledge of beef cattle genomics.

That information now sits in the Beef CRC databases at AGBU.

"The genomics database will contribute to BREEDPLAN when DNA-based prediction equations are accurate enough to provide useful outcomes for industry," Dr Graser says, noting that this is a Beef CRC objective for mid-2012.

"Early on, we thought we could use genomics to explain 50 per cent of the genetic variation which is the equivalent of an EBV with an accuracy of 70 per cent," Dr Graser says. "Now the aim is to explain 15 per cent of variation, which equates to an accuracy of 40 per cent. To get higher accuracies, we need more recorded animals."

"For most traits that are phenotypically recorded, we get accuracies that are higher than we are currently achieving with DNA markers 40-60 per cent for the animals we have on record."

Breedplan will use both sources of information, with genomic data providing an independent cross reference for the phenotypic information. Given the current pace of progress, Dr Graser expects that by mid-2012, Beef CRC researchers will have developed improved equations that are strong enough to provide a useful prediction value for traits determined through use of DNA.

However, he warns the value of the data will be compromised unless researchers are able to validate the prediction equations against performance in animals that are completely independent of animals in the discovery populations.

"We are coming back full circle," Dr Graser said. "At the start of the CRC's second term, it was the genotyping that was expensive. Ten years later, genotyping costs have dropped dramatically while phenotyping costs have increased due to higherlabor costs and economies of scale have gone down."

The Angus, Hereford, Charolais, Limousin and Brahman societies have established Breeding Information Nucleus (BINs) herds which aim to amass phenotypic data. DNA samples from every BIN calf are being col-lected, because, Dr Graser said, the genomics revolution is only just gathering pace: the most promising developments are yet to come.

The Beef CRC is currently undertaking a bid for a five-year extension that will allow it to wrap up any outstanding research and industry delivery threads. If successful, CRC researchers will utilise the BIN herds to underpin the ongoing research activities.

Given the extremely rapid changes in genomic technologies over the past five years, Dr Graser predicts that over the next five years, DNA sequencing may become so cheap and fast that the existing SNP chips may be made redundant because the full sequence will be readily available for all animals.

If the phenotypic information is available to back up the DNA data, then it is likely that at that time, genomics will begin to exponentially accelerate genetic gain, as many had hoped it might have done by now. However, the greatest requirement to achieve the desired levels of genetic gain is accurate measurements on many more animals than are currently available.

New Traits Added to BREEDPLAN Through Beef CRC Research

The introduction of scanned carcase traits to the breeding program as a result of the BREEDPLAN validation program and Beef CRC phase one, increased the dollar profit per cow joined relative to industry recording of weights alone. The subsequent inclusion of reproduction traits and actual carcase and feed efficiency traits based on the first and second phases of Beef CRC increased genetic progress even further. As at 2008, expected profit per cow meeting this market specification was around $90 and this was increasing at over five dollars/year.

The difference between the 2008 profit value and the profit value that would have occurred in 2008 if only growth rates were available (around $60 by extrapolation) is the incremental value of the new traits added into BREEDPLAN due to CRC research.

Selecting Beef Animals for Feed Efficiency Improves Carcase Traits

Selecting beef animals for improved feed efficiency will also result in larger and leaner carcases, according to Teagasc.

Selection for improved feed efficiency will not have any unfavourable repercussions for carcass traits. And, if anything, selection for improved feed efficiency – lower food conversion ratio (FCR) and lower residual food intake (RFI) – will result in larger, leaner carcasses with better conformation.

Some good news for beef producers from Teagasc's John Crowley, who led a study to quantify the genetic relationship between feed efficiency and carcass fat (CF), carcass conformation (CC) and carcass weight (CW) in Irish beef cattle.

"Feed is the largest variable cost on farms and there is much interest in selecting for improved feed efficiency. But prior to recommending selection on any trait, the expected response to selection on other economically important traits needs to be quantified," he explained to delegates at this year's British Society of Animal Science's annual conference, held at Queen's University, Belfast.

Data for this analysis originated from two separate sources. The first data set consisted of feed intake (FI) and bodyweight (BW) records collected at the Irish bull performance test station. Different measures of feed efficiency (FE) were subsequently calculated.

The second dataset consisted of information on CC, CF and CW on 2,566,969 animals slaughtered in 30 abattoirs in Ireland during 2005, 2006, 2007 and 2008. In the present study, the EUROP classification grades were transformed to a 15-point linear scale with a carcass fat score of 1 implying low fat coverage and a carcass conformation score of 1 implying poor conformation.

Only males slaughtered between 300 and 1,200 days of age and females slaughtered between 300 and 875 days of age were retained. And only animals with Aberdeen Angus (AA), Belgian Blue (BB), Charolais (CH), Friesian (FR), Hereford (HE), Holstein (HO), Limousin (LM) and Simmental (SI) breed fractions were retained leaving 822,763 records. Also animals where less than 75 per cent of their breed fraction was known were discarded. And carcass weights of less than 150kg and greater than 550kg were also omitted.

"And we found absolute genetic correlations between the performance test measures, excluding FE traits, and the three carcass traits ranged from 0.004 to 0.33," said Mr Crowley.

Although standard errors were large, ranging from 0.10 to 0.16, they were considerably smaller than other studies with similar objectives. And most of the genetic correlations with performance traits estimated in this study were not more than two standard errors from zero with the exception of the correlation between CF and BW, so selection for increased BW is expected to have a negative effect on CF.

"With regard to the feed efficiency traits, CC and CW were both negatively correlated with FCR and this suggests that selection for improved, or lower, FCR will improve both CC and CW," explained Mr Crowley.

"Furthermore, CF and RFI were correlated indicating that selection for improved, or lower, RFI will yield leaner carcasses."

He also stressed that the standard errors of the genetic correlations were large similar to previous studies, attributable mainly to the relatively small datasets for the FE variables.

"Therefore, there may be merit in pooling data from international sources to increase the dataset size and potentially generate more precise estimates of genetic parameters," he added.

Calving Ease Proofs Rate Sires For Calving Performance

Calving ease proofs will help to reduce calving difficulties and their cost, although the heritability is low.

Having calving ease proofs available will help dairy producers to identify bulls that are genetically good (or bad) for calving performance. And this may be of particular importance when choosing bulls to use – or avoid – on heifers to reduce the number of difficult calvings and cut the costs associated with losses in production, fertility and potentially cow/calf losses.

So said the SAC's Eileen Wall, when she presented the findings of her team's project, to develop routine national calving ease evaluations for UK dairy cattle, to delegates at the 2010 British Society of Animal Science's annual conference, held at Queen's University, Belfast.

The ease of calving influences the economics of a cow/calf enterprise through increased calf death loss, increased labour and vet costs, reduced subsequent reproductive performance of the cow, potential loss of the cow, and reduced milk production. It's estimated that a slightly difficult calving costs approximately £110 and a seriously difficult calving costs between £350 and £400. "So a national calving ease evaluation would be a huge step forward for the UK dairy industry," said Dr Wall, explaining the rationale behind her work.

The project took data from UK milk recording organisations, and included producer-recorded calving ease (CE) data, as well as data collected as part of the progeny test scheme. Genetic parameters for CE were estimated, considering a direct and indirect effect.

Fixed effects in the model include herd year, month of calving, lactation number, calf sex and interaction between lactation number and calf sex. Age and percentage Holstein were fitted as linear regressions. And a random effect of service sire was fitted to estimate direct CE predicted transmitting ability (PTA) and random effect of maternal grandsire fitted for indirect CE PTA, with a covariance between the two effects also fitted.

"The data spanned 15 years, but the majority of the data falls after 1999," said Dr Wall. "Overall, 84 per cent of calvings were classed as 'easy calving'. For first calving cows a total of 74 per cent calvings were classed as 'easy' and 85 per cent of later calvings," she added.

The genetic analysis showed that the heritability for calving ease was low (0.066 and 0.040 for direct and indirect effects respectively with a genetic correlation of-0.685). And there was no evidence of a genetic trend in either direct or maternal CE PTAs.

The across country genetic correlations for the multiple-trait across country evaluations (MACE) run for direct and maternal CE were in line with other countries and suggest that UK CE PTAs would be suitable for an international MACE evaluation. The genetic correlations between countries for the direct calving trait averaged of 0.80 and ranged from 0.619 (with Hungary) to 0.944 (with Canada).

The genetic correlations between countries for the maternal calving trait was a little lower with an average of 0.69 and ranged from 0.561 (with Hungary) to 0.839 (with France). These correlations with other countries are good, particularly for such low heritability traits.

Eating Quality Obtained Through Breeding

Marker assisted selection can be used in the future to allow beef breeders to select for meat quality to meet consumer demands. The low heritabilities estimated for meat quality traits, such as juiciness and tenderness, and the obvious difficulty involved in measuring these phenotypes, limits the effectiveness of traditional quantitative breeding. And making progress with these traits using direct measurement is likely to be time-consuming and costly.

So says Dr. Gill, from Edinburgh Roslin Institute. "For these reasons, we believe that the next step in the improvement of such traits is to investigate whether associations exist between meat quality traits and genetic loci, possibly starting from candidate genes.

"If significant associations can be found, then marker-assisted selection could be implemented. And this process has the added advantage of being able to assign breeding values to live animals, so that post-slaughter scoring is not necessary."

Sensory traits are known to be important to the consumer and will influence their consumption of meat, specifically beef. But these traits are difficult to measure and often require the use of taste panels to assess the complex parameters involved in the eating experience.

Such panels are potentially a large source of measurement error, which may reduce the effectiveness of breeding programmes based on the data they generate. So the aim of Dr. Gill's study was to assess the quality of such taste panel-derived sensory traits as well as calculating genetic parameters and residual correlations for these traits along with a further set of traditional carcass quality traits.

The study examined a sample of 443 Aberdeen Angus-cross animals collected from 14 breeder–finisher farms throughout Scotland. To assess the quality of the taste panel measurements, three consistency statistics were calculated: panel-member consistency, or the extent to which an individual panel member varied in their scoring for a given trait over the period of the experiment; repeatability, or the consistency with which an individual panel member was able to score a trait on repeated samples from the same animal; and reproducibility, or the extent to which taste panel members agreed with each other when scoring a trait.

"These consistency statistics were moderately high, particularly for panel-member consistency and reproducibility, with values ranging from 0.48 to 0.81 and 0.43 to 0.73 respectively," said Dr. Gill. "Estimated heritabilities were low for most of the sensory taste-panel-evaluated traits where the maximum value was 0.16 for overall liking. Residual correlations were high between many of the closely related sensory traits, although few significant correlations were found between the carcass quality data and meat quality traits."

Young Sires Benefit From Genomic Selection

Using genomic information in beef animals increases the benefits to selection, particularly for young bulls and selection parameters that have low heritabilities such as residual feed intake.

There is a potential benefit that can be achieved from including genomic information in selection programmes in beef cattle. And this benefit is highest when selecting younger sires compared to older sires.

Just one of the findings of Carol-Anne Duthie and her team at the Scottish Agricultural College. Their work outlined the importance of the training population size and the effective population size as these constrain the potential benefit that can be achieved.

"Genomic selection is expected to be of particular benefit for traits that have low heritability and are difficult to measure or are only available late in the animals life or are sex limited," she told delegates at the 2010 British Society of Animal Science's annual conference, held at Queen's University, Belfast.

"Therefore, genomic selection may facilitate the inclusion of further traits, such as residual feed intake, in breeding programmes which are important to the efficiency of beef cattle."

Genomic selection (GS) utilises information about the association of large numbers of single nucleotide polymorphism (SNP) markers located throughout the genome with phenotypic information such as growth rate. This has become feasible due to the availability of large numbers of SNP markers and the development of the bovine SNP chip. "And the aim of our study was to investigate the benefits of applying genomic selection to the terminal sire index of beef cattle considering training population size and different breeding structures," explained Dr. Duthie.

Selection index theory was applied to investigate the response of the beef terminal sire index under conventional selection or GS given the structure of the UK beef industry. Breeding objectives of the terminal sire index incorporate carcass weight, carcass fat score, carcass conformation score, gestation length and calving difficulty.

Currently recorded traits include birth weight, weight at 200 and 400 days, muscle score, muscle and fat depth, gestation length and calving difficulty. Parameters were obtained from UK beef genetic evaluations and information for the breeding goal traits were obtained from Amer et al. (1998). Selection strategies were derived from the structure of the beef industry, which was calculated from UK beef genetic evaluations data.

The effect of different breeding structures was investigated. This included selection based on young sires, where no progeny information is available, and selection based on older sires. Genomic information was included in the selection index model and accuracies of genomic estimated breeding values (GEBVs) are predicted based on trait heritability, number of phenotyped animals in the training population, the number of quantitative trait loci (QTL) underlying the trait and the effective population size (Ne).

"We found that, under conventional selection, the economic response is similar between the two breeding structures, however the slightly higher response (5 per cent) when selecting older sires is due to higher accuracy of breeding values due to the availability of phenotypic records on more relatives," said Dr. Duthie.

"Including genomic information increases the economic response to selection in both breeding structures, however the magnitude of the response in comparison to no GS is higher when selecting younger sires (up to 33 per cent increase) in comparison to the industry average (up to 21 per cent increase) where selection is based on approximately 40 per cent young sires, with a generation interval of less than three, and 60 per cent of older sires."

She added that the size of the training population influences the economic response that can be achieved when including genomic information, where the highest response is achieved with a training population of 5,000. "However, this is constrained by the breeding programme structure and Ne. Larger training population sizes had more impact when younger sires were selected. Furthermore, the rate of economic response was higher for Ne of 100 than 500."

Presented to the British Society of Animal Science Annual Conference, April 12 to 14, 2010, Queen's University, Belfast. Full details: Duthie C, Sawalha R, Navajas E, Roehe R and Roughsedge T: "Economic response that can be achieved from including genomic information to the terminal sire index of beef cattle."

To Improve Animals, You Must Improve Performance

Most beef producers still prefer to make selection decisions based on what an animal looks like, rather than a calculation of its genetic merit. However, this tradition may be undermining future profitability, writes Wayne Upton, Beef CRC extension specialist.

Most producers buy bulls the same way they have done for generations. Generally speaking, they will buy the biggest, fattest animal at a sale, and use their experience as third or fourth generation cattlemen or women to visually assess the genetic worth of an animal by what it looks like.

While tradition plays its part in keeping the status quo, another factor is that many commercial breeders do not have a strong understanding of how to use Estimated Breeding Values (EBVs) to improve the profitability of their selection choices.

They are unsure what the difference between tools like EBVs, which have been around for some 30 years, and genomics (the newer DNA markers and panels) are. For many, genetics is 'gobbledegook'.

EBVs are calculated from information on the animal's performance and the performance of its relatives and progeny. This information is used to make a prediction of the genetic worth of an animal. The newer genomic DNA markers are developed from an analysis of the minute differences in the genomic sequences of different animals which may point to genes that impact on a desired trait. In the future it is planned that this DNA information will also be used to increase the accuracy of EBVs. Although commercial breeders would like to have data on how well bulls perform for certain traits, they don't demand this information.

An example of some of the outcomes of selecting for traits using EBVs is profiled in the Southern Regional Combinations project. This project involved three sets of Angus bulls: one group selected on EBVs for intramuscular fat (IMF), the second for EBVs for retail beef yield. The third group was selected for progress in both traits. These bulls were mated to a random sample of Hereford cows and the progeny measured for IMF and retail beef yield at slaughter.

The results demonstrated:

1. Clear responses to selection of sires on EBVs for specific traits.
2. Excellent prediction of effects on carcase traits in the progeny.
3. Producers can select to improve more than one trait at the same time (even if they are negatively correlated).
4. Responses were consistent across a wide range of environments.

Collecting measurements on individual animals and their progeny is often raised as a significant issue for the northern beef industry, where cattle run in extensive production systems and where producers have managed herds over generations without measuring the genetics of animals. This is not to say it can't be done. In these extensive breeder regions, a renewed focus on heifer management, breeder performance and bull selection based on inherent fertility is imperative. The beef industry is not taking as much advantage of genetic improvement as other livestock industries, although it s at a similar level to the wool industry and the lamb industry is making improvement.

Basic Genetics

Genetics is used in the cattle industry to get a desired trait, for example-carcase yield or feed efficiency. This trait is influenced by two things.

Firstly the genetic makeup of the animal, but equally importantly, by the environment in which it is reared.

It is safe to say that each animal receives half of its genes from its sire, and the other half from its dam. However the random combination of these genes, means there is a widespread difference in the offspring. Some offspring may be exceptional, some average and some below average.

There are *additive* genetics and *non-additive* genetics, Mr Bullock told the audience of cattle producers from across the US.

Additive: These are predictable effects that get passed on from generation to generation.

***Non-additive*:** These are dominance and genetic interactions, and are influenced by how the genes match up.

Crossbreeding has occurred for centuries. By crossbreeding, producers can give the offspring an advantage over the average of the parent breeds, because of non-additive genes and the way they match up. This advantage is known as heterosis.

There are two types of traits, qualitative and quantitative.

Qualitative traits are usually controlled by one pair of genes, and are not easily influenced by the environment, said Mr Bullock.

An example of a qualitative trait would be coat colour. The coat colour of the animal will be determined by the dominant/or recessive genes carried by its parents.

For example, black (B) is dominant to red (b). Animals with genes BB would be a homozygous dominant. Animals with Bb would be known as a carrier, or a heterozygous.

Red coated animals would be homozygous recessive-bb.

If two carriers mated, there is a 25 per cent chance that the offspring would be homozygous recessive, and 25 per cent chance it would be homozygous dominant. There is also a 50 per cent chance it would be a carrier.

However, if a homozygous dominant was bred with a carrier, there is a 50 per cent possibility that the offspring will be dominant (BB), and 50 per cent chance it will be a carrier (Bb).

Quantitative traits are controlled by many gene pairs, and the environment will play a major role in the outcome, said Mr Bullock. Quantitative traits are measured by heritability, the proportion of the trait which is controlled by additive gene action.

The majority of production traits are quantitative traits, said Mr Bullock.

Because of the numerous possibilities with quantitative traits, due to environmental affects, they can be difficult to improve genetically. More and more research is being done to improve these traits, as they are often economically significant ones such as carcase yields, fertility, feed efficiency etc.

Heritability measures the proportion of a trait that is controlled by additive gene action. It is the resemblance between parents and offspring, for a particular trait, due to genetics.

If a trait has low heritability (e.g. reproduction) this indicates that there is a low involvement of additive genes and/or that the environment has a much larger influence on the trait. High heritability indicates that additive genes play a relatively large role in the trait change, and so these traits are less affected by the environment.

Whilst fertility does have low heritability, traits such as birth and weaning weight, carry high heritability, as does yield per cent of carcase weight. Mr Bullock says that the heritability of the traits does vary between breeds.

Selection

Selection is the process of introducing the most desirable genotypes that will leave offspring that are beneficial to the herd and enhance business productivity and profitability. After several generations of selection for certain traits the more desirable genes become more frequent and the less desirable genes become rarer with the result that the overall genetic and phenotype merit of the herd increases.

Selection method indexes, such as estimated breeding values (EBVs/EPDs) can result in rapid genetic improvement. Traits can be prioritised depending on economic performance, and single traits can be compared. The animals selected for breeding are those with the highest scores based on prioritised traits.

Whilst this is a reliable methods, visual assessment is still required, as not all traits can be measured.

Over time and before selection indexes, selection methods have definitely made a difference, said Mr Bullock. Looking across the breeds, there have been drastic changes in the genetic merit of most traits. This, Mr Bullock says, indicates that simultaneous selection for antagonistic traits is possible.

Measuring Genetic Merits

Measuring genetic merits is a complex task. There are number of different methods to help selection, including visual, estimated progeny differences, genomics and selection indices.

A trait is measured by taking the genetic influences and adding the environmental factors. However there is always an unexplained variation, said Mr Bullock. This variation cannot be measured. For example, if breeding with a bull that was born at a weight 75 lbs, his progeny's birth weight could vary between 55 lbs and 95 lbs.

The more genetic data available, the less the unexplained variation will be.

Concluding, Mr Bullock said that selection should be made on traits that economically impact individual herds.

What is An EBV and How can it Help You?

Estimated breeding values (EBVs) are an essential selection tool, which are underused in the UK, says William Haires, Nuffield scholar and Pedigree/Commercial beef breeder.

What is an EBV?

An EBV is a value which expresses the difference (+ or-) between an individual animal and the herd or breed benchmark to which the animal is being compared. EBVs are reported in terms of actual product e.g. days, kg of weight or mm of fat depth, etc.

There are two main performance recording services operating in the UK.

- The Australian Breedplan programme is used by many beef breeds in the UK and operates through Pedigree Cattle Services.
- Signet/BASCO is operated by Edinburgh Genetic Evaluation Services (EGENES).

Both systems produce EBVs for many of the same traits, with their databases being accessed via the respective breed societies. Sale catalogues that have EBVs included for each animal should also have an explanation of the traits and a chart showing percentile bands, depicting where the individual animal ranks within the breed depending on its performance for the traits in question.

Contrary to common belief EBVs are not a massive waste of time and money but an estimation of how the progeny of an individual should perform! The purpose of EBVs are to bring all the cattle within a herd or breed onto a common platform so that those making selection decisions can know they are comparing like with like regardless of the system or location they are produced in. It is important at this point to remember that each parent contributes half the resulting calf's genetic make-up.

This performance estimation measures the parts of the animal that the eye. An animal's physical appearance (phenotype) is determined by two components, its genetics (genotype) and non genetic or "Environmental" influences. EBVs measure this genetic component and allow cattle within their respective breeds to be compared, excluding their management and Environment. The key point here is what you see is not necessarily what you get!

The environmental (non-genetic) influences include grass quality, disease burden, parasites, supplementary feeding, management ability etc. The environment component is a part that herd keepers have influence over and the ability to compensate for or enhance. Despite the environmental influences, the genotype remains the same and parents do not pass on to their progeny the environmental effects that have influenced them.

EBVs draw information from a number of sources in addition to the animal's own performance. The basis is comparing individuals within a contemporary group, one to another, minimising non-genetic influences. The system also compares contemporaries in other herds where genetic linkages exist through siblings and ancestors. This is a very simple description of what is involved to outline the principle. The actual calculations are much more complex and beyond anything that a farmer needs to know or understand. Linked traits are those which share common connections with each other. An example of this is that where 600 day weight increases so will birth weight.

Heritability is the term used to describe how strongly a characteristic is transferred from parents to progeny and is rarely considered. It is the proportion of an animal's production that comes from its genetics and varies between traits. The higher the value the faster genetic improvement can be made and positive benefits observed. In general terms, maternal traits are poorly inherited and growth traits are moderate to highly heritable. The old adage of cattle performance being "80 per cent feeding and 20 per cent breeding" is not strictly true, but the breeding component will determine how the non genetic component is used. At low to average levels of performance, feeding and health will have more of a contribution to an animals overall performance but for those trying to maximise returns the right genetics are vital. A typical example could be milking ability, where if the genetics for milk are poor, feeding will not make a significant difference, the cow will just get fatter.

Accuracy

No two animals are the same, so the more information that can be collected, the more one can account for the variation between individuals, and so the overall risk of using them can be reduced.

EBV data must generally reach a certain level of accuracy before it can be reported to ensure the information is as relevant as possible and truly reflects the potential performance of the animal in question. Some traits such as growth will not be reported until they reach an accuracy level of 40 per cent, which is typically at the lower range of reported accuracies (these can be up to 99 per cent). Traits which are more difficult to measure such as carcass and fertility related can be reported at much lower levels due to the limited information available, requiring the user to make a judgement decision.

Producing Good Quality EBVs

Throughout a performance recorded animal's life, measurements are taken which are used to produce its EBVs. These start with a birth weight and calving ease score. These measurements are supplemented by weights at 200 days, 400 days and 600 days as well as scrotal circumference (bulls only) and ultrasonic measurement of back-fat, rump fat and muscle depth. At weaning time, dam weights will also be collected to contribute to the mature weight EBV.

In the future, genomic data may also be incorporated for the difficult-to-measure traits such as carcass and reproductive characteristics. Disposition data is collected by some overseas breeds and a docility EBV produced to measure their temperament based on "flight time". In a study carried out in Australia it was found that "flighty" cattle were gaining 0.4kg/day less than their docile contemporaries, a substantial amount in today's challenging climate. Signet will be launching a docility EBV in 2011, leading the way in the UK.

In addition to the information collected from an individual, the data from siblings and other family members make a contribution. There are many factors involved in the production of quality EBVs but it is useful to mention a few of the factors that lead to poor quality EBVs as well. The most important principle of performance recording is that of grouping the cattle together for the comparison. This is the biggest challenge for the system to overcome and depends on accurate record keeping and submission of raw data.

Where the same sire is used on the same cows year after year or where there is only a small number in the management group there is evidently a lack of genetic diversity. This limits the amount of variation that can be accounted for.

Management for Better Quality EBVs

To produce good quality EBVs it is important that the herd has a compact calving pattern, with the cows run in reasonably large groups and that any data from that group is collected on the same day. Within this group at least two different sires should be used each year to allow the system to compare the performance levels between these sires' calves. Ideally the bulls should be a mix of a proven sire and a young or unproven sire from the recorded animal itself. This information depicts the type of genetics that an individual actually possess because it helps estimate the genetics that it will pass to its progeny. Any animal that may have been sick, fed for a show or sale, used as a sire or introduced to a new group should be identified to ensure it is not compared on a "like for like" basis with the rest of its group. In small herds the use of AI bulls will lead to rapid improvements in their EBVs and accuracy. This applies to all sizes of herds. In small herds the risk may be to use a bull retained from the herd. His use is limited to only a small number of females each year meaning the chance of getting any meaningful information is negligible. Small herd owners need to evaluate the cost of keeping a bull compared to AI, given modern heat synchronisation techniques which make the process much simpler and more accurate.

The main principle must be to record all calves, the good the bad and the ugly. Selectively recording animals will deny a true picture of what is happening in the herd. It will give a distorted base set unreasonably high or it may undervalue better sires whilst overvaluing lesser sires. Accuracy in measuring weights and correctly recording dates of birth can make a substantial difference. Estimating weights is not accurate enough and whilst not having a significant effect on the overall group (as the estimate will likely be variable), 50kgs of error either way can have a much more serious impact on the individual, adversely affecting related traits especially those with negative relationships. Accurate honest weights are vital to many processes on farm including medical treatments and pre sale checks. A weigh scale is the most important piece of equipment on a livestock farm.

Popular Misconceptions

It is often asked why a good animal does not have good EBVs, or vice versa:

- Firstly an EBV is a performance estimate or prediction for the progeny of a bull or cow and not of their own performance.

- Secondly the animal will most likely have had its environment influenced in some way. It may have been taken away from the rest of its group, been suckling another or several other cows, had creep feed, been housed earlier or weaned later or some other such practice which has compensated for its lack of genetic potential.

- Below are two of the reasons why some cattle look outstanding but their EBVs do not reflect this. The questions which should be asked at this stage are "What would this animal have looked like without additional management?" and "What will his calves look like without similar expensive management?"

- Where a calf with superior genetics suffers a setback such as disease affecting it or its mother, the opposite effect can occur. This animal still has all the potential and genetics to produce great progeny but may not look as good as it could or as good as its contemporaries. This highlights the danger of selecting by eye alone, as this animal could be inadvertently rejected.

- Animals that have been imported from overseas can suffer from poor, low accuracy EBVs. With limited or no information about its past performance, its EBVs will unfortunately reflect this until sufficient data is collected and processed, in its new home country.

Concern is often expressed about the possibility of a person exaggerating the weight of an animal to increase its EBV. This could happen but it would be unwise.

There are checks and balances within the EBV system which queries animals that are too far ahead of the group and secondly, as progeny are analysed its failure to perform to expectations will be noticeable. Animals that are outside the boundaries are flagged as "outliers" for further investigation.

Anyone under the illusion that he can beat the system and fool the process is wrong. He may get away with it for one or two calf crops but the cattle will fail to live up to their expectations with their EBVs deteriorating rapidly when they go into production. The biggest victim in this fraud is the person recording the data as he is fooling himself and destroying his reputation. Breed societies need to act aggressively in any cases detected to protect their members and the integrity of the breed.

Selection Indexes

Selection indexes take the hard work out of knowing how much emphasis should be placed on each of the available EBVs when making breeding decisions. An index gives a single EBV that reflects the value of an animal in financial terms.

Indexes allow balanced selection as they apportion the amount of selection pressure that needs to be applied for growth, maternal, carcass and fertility traits to give the most profitable herd over the long term. Typical production parameters, prices and production costs underlie each index.

The next part involves steps that combine economics with genetics. Financial values for performance measures are calculated for each breed's production and market. Using genetic theory these financial values are used to calculate appropriate weightings for the EBVs currently available.

Indices exist for different markets such as terminal index and self replacing index and provide a convenient way of speeding up a search. Choosing the highest index animal will not necessarily result in a suite of traits that will suit the herd and compromise will likely be needed so it is still essential to look at the EBV for each trait and determine if they are appropriate to ultimately satisfy the previously identified needs.

How much is Genetic Improvement Worth?

At this point it is worth using an example to show how value can be calculated using EBV data. As already mentioned Indexes are a financial value. Using the example of Dorepoll 1 10H Headliner (appendix 2) who has a Self Replacing Index (SRI) of £36.00 compared to the breed average of £24.00, there is a difference of £12.00. In a herd producing 100 calves over his breeding lifetime this equates to £1200. Compare this bull to a bull with a self replacing index of £10 and the extra value to the business immediately leaps to £2600!

Irish Cattle Breeders Federation

The Irish Cattle Breeders Federation (ICBF) has developed a product named "Herdplus". This system produces management reports for both the pedigree and commercial herd owners. EBVs and indexes are also produced for both types of herd meaning that even commercial suckler cow herds have performance recording information available for all their cattle. The system draws on information from markets, abattoirs and data collected by the farmer. Hopefully the UK will be able to develop a similar system in the years to come.

Improving Performance In Suckler Herds Through Breeding Plans

Breeding is the key to a successful future for suckler herds, said William Haires, Nuffield scholar at the British Cattle Breeders Conference. Charlotte Johnston, TheCattleSite Editor reports.

Embarking on a Nuffield scholarship in 2010, Mr Haires visited the USA, New Zealand, and Australia to find out what he had to do in order to run a successful suckler herd into the future. He specifically looked at selection techniques and performance improvements, to establish the practical benefits and application of estimated breeding values (EBVs) and Gene Marker technology.

The UK has been known as the genetics capital of the world, introducing Artificial Insemination (AI). "The industry once dared to be different," he said. "Now the UK has been left behind by nations who used to look to up to us.

"Suckler herd management must change," he stressed.

With costs of production increasing, new practices and existing ones need to be evaluated. There is no opportunity to 'get it wrong' now, especially as the value of subsidies has fallen significantly.

Judging cattle performance by eye alone, is no longer a way to run a profitable business. Improving the genetic capabilities of a herd is the cheapest and most beneficial investment a herd owner can make, Mr Haires said.

He criticised the industry for moving towards continental and other breeds, which he says have let breeders focus on correcting faults in the breeds, rather than look at improving efficiency.

"Farmers have been trying to make their farm fit the system, instead of choosing a system to fit their farm," he said.

"Because breeders have become so busy treating the symptoms, we have forgotten what the problems ever were."

UK cattle have so much undervalued potential, said Mr Haires, and producers just aren't using the resources available. With rising feed costs, why aren't producers establishing breeding programmes that allow for greater use of grass-our biggest asset, he asked.

"Ultimately we want to breed cattle that perform well, are efficient and improve profitability."

So What Makes a Good Breeding Programme?

Through travelling abroad, Mr Haires has seen the value that genetic improvement has on breeding cattle to work more effectively with the resources already available.

The following are what Mr Haires considers vital when creating a breeding programme:

Market requirements: Are cattle being produced for meat or breeding? Live or dead market? Opportunities exist to produce cattle for existing or new premium markets including those based on breed, provenance, eating quality or commercial replacements.

In order to understand market requirements, knowledge is key. "It is vital that better relationships within the supply chain are developed to share requirements and knowledge which benefit everyone involved," he said.

"The current lack of shared information is restricting producers' ability to supply what the retailer and consumer demand because they simply do not know what is required."

Resources available: What land is available, what systems suit this land, environmental practices in place, buildings available.

This list is not definitive-as each farm is different and every farmer needs to evaluate his own, but whatever the resources are, they will determine the types of stock that can be kept, the potential markets that can be reached and the extent to which improvements can be made.

Herd records: This is to aid identification of cattle within a herd, and will help measure and identify individual traits at a later date.

Personal and Financial Expectations

Within herd demands: Is the herd closed, what vaccinations are required, replacement rates, stocking rates, herd docility, intervention between humans/cattle etc.

Selection tools: This is important to assist with breeding herd decisions and culling options. Selection tools could be weighing scales, EBVs or other genetic markers. Mr Haires says that cattle performance must be measured.

"The old adage of cattle performance being 80 per cent feeding and 20 per cent breeding is not true," he said. The breeding component will determine how the non genetic component is used, when trying to maximise returns, the right genetics are vital.

EBVs

EBVs should be the minimum criteria used for selecting breeding bulls, semen for AI and pedigree females for breeding, Mr Haires said.

Mr Haires found that the most successful herds he visited abroad were ones that had completely embraced EBVs. He said that whilst getting the EBV for the bull right, the bull is only half the story. Therefore cows must also be selected on EBVs.

He found that the US were struggling to get used to EBVs (called EPDs in the US), but that progressive farmers were using them successfully.

An estimated breeding value, is exactly that, he said. It is an estimation or prediction for the animals progeny, not its own performance.

Mr Haires suggests that EBV targets should be set with the system that suits the farm in mind. Bulls/cows should then be looked at and it is vital that they fit the criteria set. This way a package of traits is acquired.

Using animals with higher indexes than the average for that herd will have significant benefits for the herd, increasing the value of the business.

DNA Based Genetics

Incredible process is being made with regard to genomics. Whilst the bovine genome has been mapped out, there is a vast amount of research still needed to identify what each single nucleotide polymorphism (SNP) (found in the DNA) controls and influences.

The use of genetic markers, allows desirable traits to be recognised in cattle at one day old. Their greatest use comes with traits that have a low heritability, are difficult to measure, cannot be measured until the animal has produced progeny (by which time it may be too late), where the animal may have to be slaughtered or where the trait is not routinely measured.

"If the trait can be measured using conventional performance recording there is no need to use anything else. If there is no other method of measuring a trait, cost is not prohibitive and it is a benefit to the business, it may be worth considering genomic analysis as part of a comprehensively designed plan," said Mr Haires. Mr Haires farms 220 acres near Belfast, Northern Ireland. The farm runs pedigree Hereford and Angus herds, and a commercial suckler herd based on Ayrshire/Friesian X Hereford cows mated to native breed sires (Aberdeen Angus).

Mr Haires is currently looking at shortening the calving time, which is currently eight weeks. By shortening this, he hopes to be able to turn out animals earlier, at the start of May.

In the future, he is also looking at introducing a whole farm approach, including cell grazing. This, he says, will tighten control of the cows grazing, so they would eat areas that would usually be neglected.

One intensive grassland system that he saw in New Zealand was achieving a daily liveweight gain of 2kg per animal.

Concluding Mr Haires said: "The UK has some of the best conditions in the world to produce beef, with the use of EBV's and genomics, the future for suckler beef production is an exciting and profitable one."

Beef Breed has a Considerable Effect on End Value

Breed and sex differences in carcase distribution can have a considerable effect on the overall value of beef animals due to differential pricing of various cuts, according to studies carried out in Edinburgh and Bristol.

Steers have a greater proportion of forequarter (shin and clod) and lower proportion of hindquarter (topside, rump and sirloin) in their carcases compared with heifers.

Just one of the findings of a study, carried out by scientists in Bristol and Edinburgh, to quantify the proportions of various primal cuts within beef carcases from both Aberdeen Angus cross (AAx) and Limousin cross (LIMx) steers and heifers slaughtered through a commercial abattoir.

"This may be a function of circulating hormone levels," the Scottish Agricultural College's Jimmy Hyslop told delegates at this year's British Society of Animal Science's annual conference, held at Queen's University, Belfast.

Carcass balance, or the proportion of overall beef carcase weight that is present in various primal cuts in both the hindquarter (HQ) and forequarter (FQ) segments, has a considerable effect on the commercial value of each carcass.

So Dr. Hyslop's team took nine AAx steers and seven each of AAx heifers, LIMx steers and LIMx heifers from a range of dam types. One side of the slaughtered carcases were cut into a total of eleven commercial primals, vacuum packed and frozen at-20°C.

Although the 11 primals were further sub-divided for other experimental procedures, the weight (kg) of each commercial primal was then expressed as a proportion of the total matured carcase side weight (g/kg carcase weight) for the purpose of the 'carcase balance' study.

"And we found that LIMx animals had a higher proportion of hindquarter compared with AAx animals with the differences being larger between LIMx and AAx steers compared with the difference between LIMx and AAx heifers," said Dr. Hyslop.

"Proportional differences among 'carcase balance' can have a considerable effect on the overall value of animal breed types and sexes due to differential pricing of the various primals," he added.

Production and/or Profit?

Speaking at the recent Beef Improvement Federation Research Symposium and Annual Meeting, held in Columbia, Matt Spangler, from the University of Nebraska-Lincoln looks at focusing breeding objectives by selecting for profitable genetics, not high production genetics.

Steep increasing genetic trends for growth traits (weaning and yearling) and mature cow weight can be seen in many breeds but perhaps more alarming are those producers that have dramatically increased the genetic potential for milk production in their cow herds.

Conditional on the assumption that the Beef Cattle Industry is a For Profit organisation, then it would seem logical that profit (Revenue – Expenses) should drive our selection decisions. In order to actually do this, knowledge of environmental constraints, genetic antagonisms, and the selection tools that have the potential to measure profit are critical.

Environmental Constraints

The development of an obtainable breeding objective begins by clearly identifying environmental constraints and marketing goals.

If feed resources are limited in a stressful environment then selection for increased extreme output (high growth, milk, and red meat yield) could have negative impacts on the ability of cows to be successful breeders without the need for large quantities of harvested forage. The beginning of a profitable breeding objective is identifying what the environment will allow you to produce, at least until we have tools to apply direction selection to traits of adaptation.

Crossbreeding

Commercial producers who have not yet adopted crossbreeding are a burden to the beef industry.

We know that the two primary benefits of crossbreeding are complementing the strengths of two or more breeds and heterosis, neither of which create trait maximums. If we think about it simplistically, crossbreeding for a trait like weaning weight leaves us with a calf crop that is better than the average of the parental lines, not better than both parental lines.

Crossbreeding, if done correctly, seeks to optimise many traits through complementing breed strengths and produce animals that are better than the average of the parental lines that created them. The best tool that the commercial cattleman ever had is based on optimisation, not the production of extremes. So, it would stand to reason that within breed selection should have the same goal, optimums and not maximums.

Genetic Correlations

Unfortunately, all traits that might be included in a breeding objective are not independent of each other. Sometimes this is beneficial as we see a favourable correlated response, and other times these genetic correlations pit revenue against cost. A good example of this comes from the suite of weight traits. Depending on the targeted marketing endpoint either weaning weight (WW), yearling weight (YW) or carcase weight (CW) become a source of revenue and all are related to a major factor influencing the cost of production, mature cow weight (MW).

Although it is not intuitive, literature results show that of the immature traits, WW has the highest genetic correlation with mature cow weight.

Care should be given not to focus solely on the revenue portion, sale weight, but rather optimising input costs associated with mature weight and revenue sources from calf sale weight. The mature sale weight, CW, shows a strong and positive relationship with MW and again care should be taken to optimise selection between the two.

Selection for Decreased Input

In order to mitigate genetic antagonisms in an effort to select for profit, economic index values become the tool of choice. A bio-economic index (H) is simply a collection of EPDs that are relevant to a particular breeding objective, whereby each EPD is multiplied by an associated economic weight.

A high index value does not necessarily mean that an animal excels in all EPD categories given that superiority in trait can compensate for inferiority in other traits depending on how the EPDs are weighted in the index. A high index value should be thought of as excelling in the ability to meet a breeding objective.

The majority of economic index values are rigid (i.e. not catered to individual enterprises). A much more desirable method would use individualised index values where the bull with the highest index value may differ from one herd to the next, depending on how the animal fits the specific needs of each enterprise. While this would lead to more accurate identification of parents for the next generation, it becomes a challenging metric to use for advertisement purposes in the seedstock industry, which probably explains why this more desirable method of multiple-trait selection has not been exploited by the majority of breed associations.

New and Improved Tools

Genomic tools hold the potential to provide predictions for hard to measure traits that focus on input costs such as feed intake. Ideally, genomic predictions for feed intake would be incorporated into an economic index as a key component of input cost. However, accurate genomic predictions will require phenotypes.

The improvement of existing phenotypic databases for traits is also needed. It is critical that seedstock producers routinely turn in mature cow weights along with body condition scores to further aid in selecting for optimal weights and the development of tools such as the American Angus Associations Cow Energy value index ($EN) and the Red Angus Associations Maintenance Energy (ME) EPD. This will require participation in Whole (or Total) Herd reporting, a very necessary process for complete data collection and the development and delivery of genetic prediction tools. Trends are rarely flat, as an industry we have measured ourselves by steep lines in one direction or the other. From a seedstock perspective this may have been perceived as necessary in order to differentiate themselves (either as breeders or as breeds) from others in the market place.

Clearly identifying your production environment and realistic production goals given that environment are critical. Profit lies in the optimisation of expense and revenue and optimisation is always more challenging than maximising outputs or minimising inputs. It will require more effort, detailed financial records, and a structured breeding objective that builds a cow herd based on optimum values and not extremes. One final thought, extremely low maintenance cows will push the lower threshold of what is biologically possible for weight and produce virtually no milk. High output cows will represent the other extreme, weigh more than most mature bulls and milk heavier than the best Holstein. Both excel in some measure of the profit equation (i.e. lowest cost or highest revenue) but neither promises to be profitable.

Genomics to Improve Your Business

Today's beef cattle are the result of selection processes that have been refined for centuries. Producers have developed a strong genetic base and are designing cattle to meet the production criteria needed to fit their individual environments, says North Dakota State University beef specialist, Kris Ringwall.

Producers simply do not round up rogue cows and calves, select a calf for harvest and then invite the neighbors over for supper. Selection processes allow producers to get a feel for the genes a bull carries without having to gamble on the bull's outward appearance.

Two selection methods are readily applied today. The first is the use of expected progeny differences (EPDs). The second is the development of the newer technologies that involve analysing DNA, commonly referred to as genomics.

EPDs help the beef industry evaluate the genetics of a bull. The outward appearance, commonly referred to as phenotype, of a bull is a product of the bull's genes and how those genes interact with the local environment. Then it certainly is reasonable to make every attempt to get the best estimate of what genes are present in each bull.

For years, the environment tended to get in the way. The nice, shiny, perhaps overconditioned bull did not always sire shiny, well-grown calves in the fall.

Producers experienced a major let down and realised money was wasted on an overinflated bull. Published EPD values for each bull are intended to avoid buying mistakes.

EPDs inform producers of the expected performance of the bull's offspring relative to other bulls in the breed. Producers literally buy this expected performance when purchasing bulls with EPD values needed for their respective breeding programmes.

The EPD concept and the ability to improve a herd are based on an understanding of accuracy expressed as a value from 0 to 1. The accuracy is an indication of the amount of data going into the EPD estimate.

Accuracy includes the history of an individual's own performance data. Accuracy also includes performance data available within the pedigree of the bull and then progeny performance of the bull's offspring.

Now the good news or better news, depending how one wants to express it. Newer EPD estimates now have the capacity to add genomic information based on the analysis of the actual DNA of a bull.

The industry has been buzzing about these new analytic procedures, but a lot of chitchat would imply that the clarity of what a DNA analysis might add to a producer's beef cattle breeding program is unclear.

Producers generally grasp the concept that if a particular gene has been identified as bad and the makeup of the gene is known at the molecular or DNA level, then samples can be submitted and evaluated to see if the gene is present.

If it is, the animal can be culled. If not, the bull would not have the trait to pass on.

The same is true for published EPD traits. As mentioned earlier, EPDs are estimates that utilize the data of a bull's DNA.

This DNA, combined with management and environmental effects, is responsible for the phenotypic appearance of a bull.

The current excitement stems from newer genomic techniques that more directly link the phenotype of a bull to those long strands of DNA that stay very well hidden within chromosomes. These long strands of DNA ultimately are passed on at conception to create new calves. No, we are not necessarily creating more traits. Through genomics, we are adding to the knowledge we gain from the current EPDs available on the bulls we buy.

And that leads to increased accuracy in our selection. And that leads to better placed money when we buy bulls. And that leads to better performing cattle within the environments we are raising them in.

And that leads to more opportunity to make money. And that leads to a better beef business. And that leads to making us all happy!

Maximise Value of 2011 Calf Crop

Producers soon will be weaning calves and finalising marketing decisions. North Dakota State University Extension Service beef cattle specialist, Carl Dahlen, discusses how producers can capture the most value from their calf crop.

"For producers to capture the most value from their calf crop, they need to follow a carefully planned marketing strategy," says Carl Dahlen.

Here are four strategies to maximise calf crop value this year:

- Start with a uniform group of calves.
- Implement a sound herd health program.
- Consider age and source verification.
- Talk with your auction market representatives.

The sale price almost always is greater for a large group of high-quality, uniform calves than similar-quality calves brought to the livestock auction in smaller groups or similar calves with more weight range variation.

High-quality genetics and groups that are uniform in colour, frame, muscling, flesh and weight range are essential to add value to a group of cattle. To improve calf uniformity, concentrate on purchasing high-quality breeding stock, managing breeding seasons to create a tight calving window (consider a 45-day breeding season) and possibly culling late-calving cows.

"Implementing a sound herd health programme is the easiest way to add value to the calf crop, but it has to be done correctly," Mr Dahlen advises.

"Calves must receive the correct vaccines at the correct time."

Vaccines calves receive as they walk onto the truck do not have a large impact on the price buyers are willing to pay because most cattle will be revaccinated upon arrival at a feedlot. However, producers who vaccinated calves two to three weeks prior to weaning and followed all Beef Quality Assurance (BQA) procedures are more likely to be rewarded with higher prices.

The type of vaccine cattle receive also can impact the price. Cattle feeders are looking for calves that have received a modified live vaccine. Many feeders are not happy with the protection from a killed vaccine.

Recent sales have shown price swings of $5 to $8 per hundredweight for the same quality of calves, depending on vaccination status, with higher prices paid for vaccinated calves. Consult with your veterinarian about the vaccination strategy appropriate for your herd.

Even one year ago, auction market staff had not received a lot of requests for age and source verified calves or special age and source verified sales.

"However, 2011 is a different story," Mr Dahlen says.

"This year may prove to be the most beneficial year to date for producers who age and source verify their calves. As long as Japan restricts imports to less than 21 months of age, the demand for age and source verified calves will remain strong. But the process for verifying calves does not happen overnight."

Producers should be proactive, initiating and following through with the age and source verification process. They need to have records in place and start the paperwork well in advance of the sale to properly verify the age and source of calves.

Mr Dahlen recommends producers talk to a third-party certification agency to learn if their calves and records qualify for age and source verification. Similar organizations can help producers certify that their calves are suitable for natural and organic markets.

"I encourage producers to spend time evaluating production costs and returns to ensure that they are obtaining a sufficient premium when marketing under a natural or organic label, compared with managing and marketing calves in traditional markets," Mr Dahlen says.

Auction market representatives can be great resources for producers with questions about improving the value of their calf crop. The market representatives deal with questions about feed yard preference for calves, status of calf supply and demand, whether feedlots are running at capacity and the latest market trends daily, whereas many cow-calf producers evaluate these questions once a year.

In addition, auction markets may offer their customers alternative marketing avenues, such as special sales, Internet sales and video auctions.

Producers should remain in contact with auction market staff and be flexible in their marketing strategy, according to mr Dahlen. Consign cattle well in advance of the sale and let the auction market know as much about the calves as possible (for example, what colour they are, when they were born, when they were vaccinated and whether they are age and source verified). The more information the auction market has, the better job it can do of marketing those calves.

In addition, producers should ask auction market representatives about other ways to improve their cow herd and calf marketability.

"No silver bullet will make you consistently top the sale each time you sell calves," Mr Dahlen says.

"Concentrate on creating a uniform set of calves with cows that work in your environment. Manage your herd with good BQA practices, vaccinate in a timely fashion, age and source verify if the market is rewarding the effort, and remain flexible with marketing strategy and dates."

Make Sure You Know What Your Gross Margins Are

When a producer understands gross margins, he or she can start to appreciate what dollars are available to work with, says Kris Ringwall, North Dakota Beef Extension.

A point of clarification is in order. Assessing income in a cow-calf enterprise is more than just the price received for the calves. The impact of making change also needs to be measured. If the impact of the change is measured only in calf value or dollars received per pound of calf sold, the measured impact is not the most accurate.

The assessment of the health of a beef enterprise is critical. If the current managerial efforts are generating the desired outcome, pen and paper should be able to track the effort. Did net return go up and did the desired managerial effort have a positive outcome? The measuring stick that is most beneficial is the calculation of gross margin per cow. That value is not the same as the calf value per cow. That is because there are multiple sources of income within a cow-calf enterprise.

The natural and appropriate tendency is to focus on cost as one evaluates a cow enterprise. However, if a major change in management is made such that the historical income opportunities also have been modified, then both the cost and income side of the operation must be assessed properly to understand fully the results that are positive, neutral or negative.

That is why the call for a point of clarification. Gross margins per cow should be discussed as a finite value. To change the value (increasing or decreasing gross margins), the dollar change is not directly connected to a particular dollar value in terms of price per pound of calf or any other livestock sale. It is a dollar value projected by the composite sale of all cattle products. This is the overriding business principle in using gross margins.

When a producer understands gross margins, he or she can start to appreciate what dollars are available to work with. How those dollars are obtained depends on what livestock are available to sell.

According to Jerry Tuhy, Bismarck State College farm business management instructor at the Dickinson Research Extension Centre, gross margin accounts for the purchase and sale of all calves, cull cows and bulls, plus the animals transferred in and any overall changes in cattle inventory.

The North Dakota Farm Management education programme, along with FINBIN data from the Centre for Farm Financial Management at the University of Minnesota, indicate that, since the turn of the century, per cow gross margins have averaged $501.42.

To get to that number, questions based on dollars per cow in the cowherd need to be answered. The answers will reflect the FINBIN estimates for the years 2000 through 2010. What was the value of calves sold per cow ($180.96)? What was the per-cow value of the calves transferred out of the cow-calf enterprise to a backgrounding or feedlot operation ($358.85)? What was the per-cow value of cull sales that includes bulls, cows or young breeding stock ($99.30)? What other miscellaneous sales of beef were accounted for per cow ($5.74)?

As a producer, if I stop there and add up the income per cow, the total sales income per cow would be $644.85. Keep in mind that, even though this is the average value of sales per cow, the average value per pound of calf sold cannot be calculated by this number. In fact, even if one looks at only the value of calves sold and calves transferred out, the average value per pound cannot be determined by these numbers.

It is critical to think gross margin concepts instead of value per pound of calf. However, we still do not have the gross margin calculated because the cowherd needs to be replaced. We have old cows going out and new cows coming in. Therefore, what was the cost per cow of all purchased breeding stock ($77.75)? What was the cost per cow of breeding stock transferred into the breeding herd ($69.07)? Finally, did the inventory within the herd change at all? If so, what was the value of that change ($3.38)?

Right now, all the numbers for calculating gross margin are available. If the sale numbers are added up, plus inventory change, one gets a gross margin of $648.23. To offset the cost of replacing the herd, one needs to subtract the cost of purchased or transferred breeding stock. If one does that, the average gross margin per cow since the year 2000 is $501.42.

The Bayesian Controversy in Animal Breeding

Frequentist and Bayesian approaches to scientific inference in animal breeding are discussed. Routine methods in animal breeding (selection index, BLUP, ML, REML) are presented under the hypotheses of both schools of inference, and their properties are examined in both cases. The Bayesian approach is discussed in cases in which prior information is available, prior information is available under certain hypotheses, prior information is vague, and there is no prior information. Bayesian prediction of genetic values and genetic parameters are presented. Finally, the frequentist and Bayesian approaches are compared from a theoretical and a practical point of view. Some problems for which Bayesian methods can be particularly useful are discussed. Both Bayesian and frequentist schools of inference are established, and now neither of them has operational difficulties, with the exception of some complex cases. There is software available to analyse a large variety of problems from either point of view. The choice of one school or the other should be related to whether there are solutions in one school that the other does not offer, to how easily the problems are solved, and to how comfortable scientists feel with the way they convey their results.

Animal Breeding & Genetics

The animal breeding and genetics section works to provide new scientific discoveries to age-old livestock production problems to help producers and consumers.

Currently several courses are available to undergraduate students interested in animal breeding and genetics, and both master's and doctorate degrees are offered in animal breeding.

The field of animal breeding and genetics research is more exciting than ever before, with projects such as bovine gene mapping and DNA sequencing. Using state of the art tools and facilities, the researchers at TAMU are able to contribute to the field of animal biotechnology on a worldwide level.

Animal Breeding and Genetics

Farm animal breeders have two core activities. By using selection techniques, they develop improved varieties of sheep, cattle, chicken and so on. And by accelerating reproduction they either produce, or assist the farmer to produce, large numbers of offspring of these varieties. In connection with these core activities, breeders undertake ancillary research on biotechnologies such as Artificial Insemination (AI) and collect data on animal productivity and health. For much of the twentieth century, the goal of breeding was improved production. Breeding targets included high milk yield in cows and large, fast-growing muscle in chickens farmed for meat. Recently, however, the emphasis has changed. Breeding programmes now aim to produce animals that are more disease resistant, enjoy higher levels of welfare, and provide good quality food.

Ruminants (Cattle, Sheep, Goats)

The cooperatives can provide their members with semen from a wide range of breeds, such as Holstein-Friesian, Jersey, Montbéliarde or Nordic Red breeds. Breeding organisations pursue research on AI and embryotransplantation, and undertake data registration (e.g. milk-recording and family information) for their members. Breeding information on dairy bulls is collected from around the world by the International Committee for Animal Recording (ICAR) so that international comparisons can be made.

AI use in ruminants varies a great deal. In dairy cattle in the Nordic and West European countries it accounts for nearly 100% of inseminations. But in sheep and goats, and in some Southern European countries, it is rarely used.

Countries in which AI is the main method of dairy cattle insemination use between one million and seven million doses, or 'straws', of semen each per year. (Precise figures depend on the scale of dairy farming in each country.) Between 1 and 3 artificial inseminations are normally needed per conception, and pregnant cows generally carry a single calf.

AI in beef cattle varies depending on the country and the breed. Overall, however, it is used less in beef cattle than it is in dairy cattle. Where conception is achieved through natural insemination, one beef bull normally serves around 25-30 cows. In sheep and goats insemination is usually natural. In sheep, one ram serves at most 30 ewes. Embryotransplantation is used commercially in cattle. Globally, just over half a million cattle embryos were transferred in the year 2000. In total 20% of these were European and 42% were North American.

The main European users of embryotransplantation were France, Germany, the Netherlands, the UK and Italy. In 2000 a total of 440 sheep embryos were transferred in vivo in Europe. In the same year 145 goat embryos were transfered.

Poultry (Chickens, Turkeys, Duck, Geese)

A small number of privately owned poultry breeding companies provide Europe and the rest of the world with breeding stock. Indeed all the larger breeding companies in this sector supply products around the world. This situation has evolved both because poultry have a high reproductive rate, and because there is intense competition within poultry breeding and the sophisticated breeding programmes needed to survive this competition are costly.

To expand on the first of these factors, in broilers (chickens raised for meat) one million grandparent stock can produce 50 million parent stock which in turn produce 600 million birds for the market. A single breeding company can therefore supply a huge number of farms. Fish are like poultry in this respect, and cattle are not. In layers (chickens raised to lay table eggs) two companies together serve 80-95% of the European market and over 75% of the world market. Smaller companies are active in domestic and niche markets, and one North American company has a foothold in the European market. Three breeding companies dominate the European broiler market and the world market. Two companies have breeding locations in both Europe and the USA.

One large commercial company also produces non-intensive stocks, and the market in this area is growing: in France, 112 million birds bearing the non-intensive certificate, Label Rouge, were sold in 2000. However, although freerange and organic varieties of bird fetch high prices on the market, little of this revenue trickles down to the breeders, who receive only a small percentage of the mark-up.

Pigs

European pig breeders are currently enlarging their European and global markets. Improved methods of freezing semen for export make this possible. Unlike the situation with cattle, where frozen semen can be used without affecting fertilisation adversely, this results in more conceptions per insemination.

AI usage in pigs differs enormously from country to country, ranging from over 90% of inseminations to virtually none. Embryotransplantation is not yet used commercially in pigs. Within Europe, around 2,500 embryos were transferred in vivo during 2000, but most of these were either used in research or stored to preserve biodiversity.

Fish

The genetic improvement of farmed fish has developed rapidly since the mid-1980s, and nowadays over 70% of EU aquacultural production (especially salmon, rainbow trout and turbot) is from selected stocks. Investments during the 1990s lead to the domestication and selective breeding of sea bass, sea bream and African catfish. Trials of other aquatic species, such as cod, halibut and tuna, are also underway.

From the breeder's perspective, fish are rather like poultry. They are relatively small, produce a large-indeed enormous-number of eggs at a single spawning (between 2,000 and 50,000), and are normally reproduced through artificial fertilisation. As a result of these similarities, the fish breeding sector tends to be evolving along the same lines as the poultry breeding sector, with a small number of companies becoming dominant. In aquaculture, unlike poultry, breeding is often integrated with other activities-e.g. with the husbandry, slaughtering, processing and selling of the fish to the supermarkets.

In France, 16 companies with interests in oysters, rainbow trout, brown trout, brook trout, sea bass, sea bream and turbot combine their research and jointly evaluate their breeding programmes.

The reproduction of farmed fish involves eggs and semen being gently stripped from adult fish by hand. The semen is then introduced into water containing the eggs, and fertilisation takes place very much as it would naturally. Like most plants consumed today, farmed fish may have more than two sets of chromosomes.

This does not mean that they are GM fish. It merely means that, as 'sterile triploids', they are unable breed with their wild relatives when, occasionally, they escape. The practice of inducing triploidy is recommended by international organisations such as the FAO, ICEM and NASCO.

6

Economics of Dairy Beef Production Systems

Introduction

With the impending abolition of milk quotas in 2015, it is projected that dairy cow numbers in Ireland will increase substantially. This will result in a greater number of dairy male calves becoming available for beef production. Although the majority of beef cattle from the dairy herd are currently finished as steers, there is increasing interest in bull finishing systems given the inherent greater live weight gain performance of bulls relative to steers. However, these systems have typically involved greater levels of concentrate feeding than steer systems. The objective of this paper is to examine the economics of a range of options for finishing male calves from the dairy herd. These systems are based on the systems currently under evaluation in the Johnstown Castle dairy calf-to-beef project. Six options were evaluated representing calf-to-beef systems finishing male calves as; veal at 8 months of age, bull beef at 12, 16, 19 and 22 months of age and steer beef at 24 months of age.

Veal Finishing Systems

Veal finishing systems using Friesian and Jersey crossbred calves were evaluated. In this system calves were finished on ad libitum concentrate diets with straw offered as a source of roughage following a 12 week rearing phase. Live weight gain was lower for the Jersey crossbred calves resulting in a 20 kg differential in slaughter weight and a 14 kg differential in carcass weight in favour of the Friesian calves. Approximately 750 kg of concentrates were fed per head. For the financial analysis of the other systems evaluated, cost and price assumptions were based on those prevailing for the Johnstown Castle dairy calf-to-beef project in 2011. Jersey crossbred calves were purchased at €30/head with Friesian calves costing €140/head. Fixed asset requirements were assumed to consist of a calf house and a weanling finishing unit. Both of these were assumed to be in the seventh year of a 20-year life span with interest charged at seven per cent per annum. A similar approach was taken for the remaining production systems evaluated in this paper with finishing housing costs reflecting age and weight at finish. Both Friesian and Jersey crossbred calves returned a positive net margin. Systems based on Jersey crossbred calves were somewhat more profitable largely owing to lower calf purchase price and lower concentrate feed requirements. Sensitivity analysis indicated that both systems are very sensitive to calf price and veal price in particular.

Twelve Month Bull Finishing System

The 12 month bull finishing system and the remaining beef systems evaluated in this paper, had a similar rearing phase to the veal production system. Following the rearing phase calves were built up onto ad libitum concentrate with straw offered as a source of roughage. Similar to the veal production system, live weight gain was greater for the Friesian compared to the Jersey crossbred calves such that slaughter weight was 64 kg greater and carcass weight was 40 kg greater for the Friesian bulls. Correspondingly, concentrate consumption was greater for the Friesians. Neither system returned a positive net margin. However, gross margin was positive for both systems indicating that a contribution can be made to the fixed costs of the farm. In addition to calf price and beef price, the 12 month bull system is also sensitive to concentrate price.

Under 16 Month Bull Finishing Systems

For the under 16 month bull finishing system, calves were turned out to pasture in May at approximately three months of age for a six month grazing season. Two treatments were compared; supplementing at pasture with 2 kg of concentrate (PC) or offering no supplementation during the grazing season (PO). Following housing in early November, all cattle were adapted onto an ad libitum concentrate diet with straw offered as a source of roughage. Cattle remained on ad libitum concentrates for ~200 days before slaughter at under 16 months of age. Friesian bulls were ~50 kg heavier at slaughter than Jersey crossbred bulls and cattle supplemented during the grazing season

(PC) were ~45 kg heavier than cattle offered pasture only (PO). Total concentrate supplementation ranged from 2.3 t for Friesians supplemented during the grazing season to 1.5 t for Jersey crossbreds offered pasture only during the grazing season. Financial results indicated that for systems finishing bulls at under 16 months of age, Jersey crossbreds were more profitable than Friesians and pasture only was more profitable than supplementation during the grazing season. All systems returned a positive gross margin ranging from €70/head for the Friesian system where calves are supplemented in the first grazing season to €194/head for Jersey crossbred systems where calves are not supplemented in the first grazing season. However, when full fixed costs are allocated to these systems, only the latter system (Je PO) returned a positive net margin.

Nineteen and Twenty Month Bull Finishing Systems

These systems operated similar to the under 16 month bull finishing system up until housing in November at the end of the first grazing season. Thus, two treatments were imposed during the first grazing season; calves supplemented with 2 kg concentrate (PC) and calves receiving pasture only (PO). Following housing, bulls were fed ad libitum grass silage plus 1.5 kg of concentrate until turnout to pasture for a second grazing season in early March. The bulls for finishing at 19 months of age were housed in mid May for a 100 day finishing period on ad libitum concentrates with straw offered as a source of roughage.

Similarly bulls to be finished at 22 months were housed approximately 3 months later for a similar 100 day finishing period. Bulls finished at 22 months of age were 70 kg and 90 kg heavier for PC and PO groups, respectively, when compared to bulls finished at 19 months of age. Total concentrate intake over the lifetime of the animal ranged from 1.4 t for 19 month finishing bulls which were not supplemented during the first grazing season to 1.8 t for 22 month bulls receiving supplementation during the first grazing season. All systems returned positive gross and net margins with net margin ranging from €150/head to €265/head for 19 month bulls not supplemented in the first grazing season and 22 month bulls supplemented in the first grazing season, respectively. Twenty-two month bull systems were more profitable than 19 month bull systems and systems where concentrate supplementation is provided at pasture during the first grazing season were more profitable than systems where no supplementation was provided during the first grazing season. All systems were highly sensitive to beef price in particular and concentrate price. Again, calf price was also important.

Twenty Four Month Steer Finishing Systems

In this system Friesian calves are finished for slaughter as steers at 24 months of age. Calves spend their first season at pasture with no supplementation. Following an indoor winter period during which grass silage is offered ad libitum in addition to ~ 1 kg/day concentrate supplementation, yearlings are turned out for a second season at pasture. Cattle are housed in October/November for a ~150 day finishing period. During the finishing period good quality grass silage was offered in addition to 5-6 kg concentrate daily. Target slaughter and carcass weights were 620 kg and 320 kg, respectively. Total concentrate intake was ~1.2 t. Gross and net margins for this system were €355/head and €193/head, respectively. Margins were most sensitive to beef price with concentrate price and calf price assuming similar sensitivity.

All of the systems evaluated in this paper returned positive gross margins indicating that these systems provide a contribution to fixed costs. The 22 month bull finishing system was most profitable with the 24 month steer system next most profitable.

The veal production system was also competitive with these systems arising from the much higher price received for veal meat. However, this system is for niche producers with a relatively small market demand. However, for the 12 month and under 16 month systems gross margins were insufficient to cover allocated fixed costs and therefore, these systems returned negative net margins. The exception was the under 16 month Jersey crossbred system where calves were not supplemented during the first grazing season which had a positive net margin. It is apparent that the economics of dairy beef systems are highly sensitive to beef price, concentrate price and calf price and thus, the market outlook is of critical importance when evaluating the profitability of these alternative systems. Thus, it is recommended that detailed enterprise budgets, subject to the prevailing conditions on individual farms and including sensitivity to key parameters, are prepared annually on dairy calf-to-beef enterprises. It should be noted that a constant beef price has been assumed for all scenarios. As the cattle in the Johnstown Castle research project are slaughter, and carcass data becomes available, it will be possible to re-evaluate the economics using the price received for the alternative systems.

Suckler Beef Production: Animal Productivity

Animal Productivity

To ensure high levels of profitability from suckler beef production systems, animal productivity, or output per livestock unit, must be high. High output per LU is determined by weight for age and carcass quality of the progeny and reproductive performance of the suckler cow herd.

Weight for Age and Carcass Quality

Drennan and McGee (2009) identified three important factors influencing growth rate and carcass quality of suckler progeny: 1) use of late-maturing continental breeds, 2) availing of hybrid vigour and, 3) milk production of the dam. Drennan and McGee (2009) concluded that suckler dams should have at least 50 per cent and preferably 75 per cent of a late-maturing continental breed to produce progeny suitable for higher-priced markets as a result of improved conformation and leaner carcasses. Murphy et al. (2008a,b) found that progeny from crossbred cows with Friesian or Simmental ancestry had higher carcass weight for age than ¾ or purebred beef breed suckler cows. These differences in carcass growth reflected differences in calf pre-weaning gain due to milk yield of the dam. However, progeny from cows with Friesian ancestry had poorer conformation and were fatter than those from purebred beef breed cows.

The implications of weight for age were investigated in a whole farm systems context to elucidate the impact of this single variable on beef output for suckler beef system. This analysis suggests that weight for age is a key determinant of carcass output and hence profitability. The implications of this for whole farm profitability were that, for each 25 g/d increase in live weight, net farm margin was increased by €30/ha. Thus, it is apparent that weight for age (and also taking into account carcass traits, as the final product must be commercially saleable and of high value), is an important factor determining profitability for suckler beef production systems.

Data from the Irish Cattle Breeding Federation (ICBF) suggests that average calving rate (i.e. number of live calves produced per cow on the farm) for Irish suckler beef farms is 0.81. In other words, for every 100 cows, only 81 weanlings are produced. This low level of reproductive performance is an obvious contributor to low levels of profitability on Irish suckler beef farms as the cost of carrying each suckler cow is only offset by 0.81 weanlings.

The calving rate is largely a function of two variables; calving interval and pregnancy rate. Analysis of data from spring-calving suckler cows at Grange between 1987 and 1999 showed that a calving interval of 367 days and a pregnancy rate of 94 per cent were achieved (Drennan and Berry, 2006). This analysis also showed that earlier calving cows had longer calving intervals than those calving later in the spring, however, there was no difference in calving rate. The results show that under appropriate levels of management, good reproductive performance can be attained in a spring-calving suckler herd. As a result of higher calving rates and higher weight for age, output per LU is greater for the Grange systems. Target output per LU is 50 per cent greater for the Derrypatrick Herd when compared to the current sectoral average. It is anticipated that improvements in reproductive performance and animal liveweight gain could increase output per LU by 11 kg by 2018. When these levels of individual animal performance are allied to much higher stocking rates, it is evident that live weight and carcass weight output is also much greater.

Minimising The Stress of Weaning Beef Calves

Weaning of calves is usually done abrupty and early compared to the natural weaning of the species. This causes social and environmental stress to calves.

Main Stressors Associated With Weaning

At weaning beef calves are usually submitted to the sudden and simultaneous loss of the social contact with the dam and the milk she provided. The former involves the loss of access to the udder and thus suckling behaviour, and a break of the bond with the mother. Usually, weaning also involves changes in social and physical environment. Although these stressors can be presented separately, what normally happens is a superimposition of several or even all of them. Thus, it is difficult to discuss the relative importance of the loss of milk, the loss of the udder and the separation from its dam for the beef calf.

The motivation to maintain the social bond by both parties goes beyond obtaining milk, because besides nutrition nursing also provides emotional comfort to the young. Nonetheless, six-month-old beef calves that were prevented from nursing with the use of nose-flaps displayed increased vocalization, walking and reduced playing, ruminating and grazing, indicating that cessation of nursing may contribute to the weaning distress response in beef calves even at this age.

Milk is a food rich in protein and energy, and in the case of beef cattle it has been estimated that the amount of milk produced by cows six to seven months after birth can provide approximately 30 per cent of the metabolizable energy required for European breeds of calves raised on pasture. Some studies reported a decrease in growth rate and even weight loss in beef calves weaned at around six months, whereas in others no change in growth was found. Weaning distress appears to be greater in calves that suckle cows with higher milk yield and are heavier at weaning, but not when live weight is controlled. Thus, differences are likely related to the development of the calf, and the amount and quality of solid food available before and after weaning in different studies.

In pasture systems it is common to move weaned calves to new paddocks, changing the physical environment, and possibly influencing the response to weaning. Though we are not aware of studies in beef cattle, it has been shown in studies with deer calves (Cervus elaphus), foalsand piglets that remaining in the same place after weaning reduces the effects of weaning stress. In confinement systems, animals are also usually subjected to new housing and diet, for example, changes from pasture or hay to concentrate feed. Similarly in pasture-based systems it is quite common to isolate the calves during weaning for about a day in a corral, and then move them to a new paddock. In most cases, the corral or new paddock may be environments totally unknown to the calves; thus, they do not know the location of resources such as food, shade, or water source. Moreover, changing the physical environment can interfere with the animals' ability to recognize members of their group, which can generate social stress.

Although mixing of unfamiliar animals is less common when weaning beef cattle than other species, the mere fragmentation of stable groups during weaning can act as a stressor. For example, the separation of a group of cows and their calves from the main herd five days before the day of weaning was sufficient to increase the concentration of cortisol in the blood of cortisol in the blood of the calves. Further studies should consider the effect of social disruption on the overall response of beef cattle to weaning. This may be done comparing the response to weaning in groups that are submitted to social disruption and other that remain in their social group.

Physiological and Behavioural Responses to Weaning in Beef Calves

Some studies have addressed the physiological responses to weaning in calves. Abrupt weaning at six months causes increases in plasma cortisol and norepinephrine. Also, another study reported an increase in peripheral catecholamine concentration in response to separation, and a subsequent decrease when the same cows and calves were reunited. An increase in plasma cortisol and heart rate also occurred after separation of dairy calves from their foster dams at three months of age. Acute phase protein concentration also increases after weaning. An increase in the ratio of neutrophils and lymphocytes, and a reduction in antioxidant enzyme activity of leukocytes, which indicate the presence of oxidative stress, has been described in beef calves weaned at seven months. A transitory reduction in immune function, peaking between d2 and d7 after weaning, was reported in grazed beef calves abruptly weaned at seven months.

Among the many behavioural changes taken as indicators of weaning stress, perhaps the most characteristic is the high frequency of vocalizations emitted by the calf. Vocalizations by the young are thought to evoke maternal care and the need to reunite with the dam. According to evolutionary theories a honest signal must fulfil four requirements: 1) there must be a degree of relatedness between sender and receiver of the signal; 2) the emission intensity and the benefit obtained should be proportional to the need for resources by the offspring; 3) emitting the signal has a fitness cost; and, 4) the receiver (the dam) must obtain a fitness benefit by providing resources to the signaller (the offspring). Thus, the vocalizations that typically follow weaning are considered a reliable signal of the emotional and physiological condition of the calf because the energy cost and high risk of attracting predators can be compensated by the high-value resources provided by the dam. Moreover the high frequency vocalizations that are often associated with abrupt weaning may also indicate the animal's state of frustration for being unable to receive food, care or reunion with the mother.

An increase in general activity and walking frequency, and pacing have been reported in beef calves immediately after weaning. In general, weaned calves increase time they remain standing, and remain little time resting compared with pre-weaning time budgets. These behaviours, together with vocalizations, have also been interpreted as a sign of motivation to reunite with the dam. Some changes in the feeding behaviour of calves can also be observed immediately after weaning. Usually there is a reduction in the time grazing and consuming other solid foods, which is accompanied by a reduction in rumination time, probably due to a change in diet intake and composition.

Weaning also triggers a marked and progressive change in the general behaviour repertoire of calves, which starts to adopt a behaviour pattern more typical of adult cattle, characterized by the predominance of maintenance activities. Whereas during early life most social interactions of calves involve the dam, after separation the relationships with the other animals in the group become more important. After weaning playing frequency is reduced and aggressive and affiliative interactions within groups of calves increase. Also, the behaviours within groups of weaned calves becomes more synchronized, with greater spatial cohesion and social interaction than among suckling calves. This change probably occurs because before weaning the behaviour of the calf is more associated to that of its dam.

Strategies Used to Reduce Weaning Distress

Different weaning methods have been used with the goal of reducing the negative consequences of weaning on behaviour, performance and welfare. Some of these methods aim to enable the calf to cope with the change in diet that accompanies the separation from the dam, while others attempt to mimic the natural weaning process, by causing the loss of milk to occur before the final separation from the dam.

Previous contact with the solid food that will be provided after weaning may result in a gradual replacement of milk with solid food while the animal is still in contact with the mother, thus encouraging independence from the mother as early as possible. Practices such as "creep feeding" or "creep grazing" in which the calf has access to feed or pasture of high quality, respectively, have been successfully used to stimulate the calf to eat solid food and thus progressively reduce its nutritional and social dependence on the cow. Although especially important in the case of early weaning, these techniques should also be encouraged when weaning is carried out at conventional ages, as it also produces positive results.

Beef calves conditioned to hay prior to weaning ate longer and had a reduced behavioural distress response at weaning compared to calves that had not had prior contact with hay. Since feeding behaviour is influenced by social facilitation and learning from peers, the inclusion in the group of animals that are familiar with the food may encourage consumption of the novel food by weanling calves, thus improving their welfare. However, although there are anecdotal reports of positive results, the few studies that evaluated the method found no benefit to the health or growth of these animals, and suggested the possibility that the presence of unfamiliar animals may cause stress.

The transition to a fully solid diet while the calf is still with the dam has been forced by the use of nose plates in calves, which prevent nursing but allow the consumption of solid food. Six to seven-month old beef calves that were weaned using nose-flaps over 5 or 14 days vocalized and walked less, had fewer agonistic interactions and spent more time eating and lying down, than calves separated abruptly from their dams when finally weaned.

Although the use of nose-flaps reduced some of the behaviours associated with stress at weaning, in some cases these animals also had lower average daily weight gain. Moreover, the welfare of calves may also be impoverished by the fact that after the placement of nose-flaps there are several attempts to nurse and the calf stays in closer proximity to the cow, which suggests that these calves may be frustrated by not having access to a resource that is otherwise apparently available.

Another weaning method involves separating the cow-calf pair through a fence for a few days before the definitive separation, which allows partial physical contact, while preventing suckling. However, results from studies investigating this method of weaning are contradictory, possibly due to different duration of the separation, timing of the observations, added to genetic and environmental variation between the studies.

In one study, calves that had been separated from their mothers through fences during a short period of time before the final separation had higher daily weight gain, spent less time walking, emitted fewer vocalizations, and spent more time eating and lying than abruptly weaned calves. In another study, beef calves had a higher frequency of behaviours indicative of stress during temporary separation from the mother through a fence, than calves that were separated but had no contact with their mothers. No benefits for weight gain or biomarkers of oxidative stress were found when calves were weaned after seven days of fenceline separation.

Calves weaned after 14 days of fenceline separation grew less than those weaned abruptly, and vocalized more frequently and over a greater number of days during the period of partial separation. Furthermore, in two of these studies during the first days of fenceline separation calves spent more than half the time near the fence that separated them from the cows, suggesting a high motivation to reunite with the dam. A similar response was reported in fence-separated foals and lambs. Thus, it is possible that the fact of being unable to fulfil a strong motivation to suckle or make physical contact with the dam may be a source of stress and frustration for these young animals.

Greater knowledge of the physiological mechanisms involved in the natural weaning compared with artificial disruption of the maternal-young bond may bring some light into the underlying mechanisms. Also, clearer understanding of the relationship between physiological changes and the resulting behaviours associated with weaning are needed, in order to assess the magnitude of the weaning distress response. This may also help guide the development of effective practices to minimize the stress associated with weaning. Most information in this area comes from studies with rodents, which may not translate well to ungulates and specifically domestic cattle. Thus, understanding the physiological changes during natural, early abrupt weaning and weaning with the alternative methods covered earlier in this review may help propose hypotheses regarding the relative contribution of the social and nutritional losses to the calf.

Although weaning is considered a major source of stress for beef calves, it is also considered a necessary practice to ensure reproductive efficiency, accelerating rebreeding of the dam postpartum, and thus increasing pregnancy rates. Therefore, as weaning practices will probably be widely applied in grazing systems, techniques to minimize weaning distress should be investigated and developed, and included in practical management.

Methods currently applied in attempt to reduce distress associated with weaning involve mimicking the gradual changes in diet and social bond of natural weaning. Studies assessing such methods have provided conflicting results, with some suggesting that step weaning using fenceline separation or nose-flaps may be beneficial, others concluding that they do not influence the outcome for the calves or that it may even impoverish welfare to some extent. Instead of reducing the magnitude of the stress caused by abrupt weaning, these practices may redistribute the response in two episodes, one when the motivation to suckle or establish full physical contact are prevented, and another one at the moment of the definitive separation.

These results lead to further questions about whether these methods actually provide an overall benefit for the calves and justify the extra management involved, such as moving animals for nose-flap fitting and verification of its permanence, or preparation of appropriate fencing to keep cows and calves apart.

Further studies should assess possible influences of milk yield, pregnancy and metabolic state of the cow during lactation, as well as the influence of food availability and quality, both before and after weaning, on the effect of these methods on the calves' growth, behaviour and physiology.

Another unexplored issue is the possible influence of previous contact with humans and habituation to handling, on the response to weaning. Weaning stress may be reduced, for example, by exposing the calves to the new environment features they will find after weaning, such as feedstuffs, location of drinking water, new social partners, humans and handling devices. Furthermore, as younger weaning ages are being increasingly recommended and adopted in pasture systems, in younger animals the sudden shift to a fully solid diet may have greater negative impact than for calves weaned at conventional ages.

Thus, the influence of age at weaning on the distress response, as well as the effectiveness of current alternative weaning methods for reducing weaning distress in younger animals need to be assessed. Studies should focus especially at the nutritional aspects associated to weaning, as these are likely to be more important at younger ages.

Suckler Beef Production

Irish grassland has the potential to produce high yields of highly digestible herbage due to favourable climate and soil types. Thus, Irish livestock farmers have a competitive advantage when compared to pigs, poultry and cattle feedlot systems, which require high quantities of concentrate feeding. Irish suckler beef production systems must exploit this opportunity to grow and utilise grass efficiently. The technical aspects of growing and utilising grass efficiently are described in the paper by O'Donovan, Hennessey and O'Riordan (this proceedings). A key objective must be to maximise the proportion of grazed grass in the annual feed budget of suckler beef systems. Turnout date of suckler cows and progeny is a critical element influencing composition of the annual feed budget. Where grass is available and where grazing conditions are appropriate, earlier turnout increases the proportion of grazed grass in the total farm feed budget and hence, improves profitability.

Advancing turnout date by one day increases net margin by €1.54 per cow. Similar to the calving date analysis, effects are due to differences in feed and slurry handling costs. Earlier turnout has also been shown to result in improved animal performance (Kyne et al., 2001; Gould et al., 2010; O'Riordan et al., 2011), although in these studies this advantage is largely diminished by the end of the grazing season due to the effects of compensatory growth.

The start of the grass growing season differs from location to location and therefore, turnout date will also vary. Whilst grazing conditions are largely dependent on soil, climatic and weather conditions and is therefore, largely outside the farmer's control, farmers can have an influence on pasture availability by appropriate autumn grassland management and judicious application of nitrogen (N) fertiliser.

Spring response to N is dependent on soil temperature (Black, 2009) and therefore, varies greatly among years and locations. O'Donovan et al. (2004) found pasture response rates ranging from 5.6 to 15.6 kg pasture per kg N applied on free-draining soils in the south of Ireland. Using this range in N response rates, the impact of turnout date and pasture N response rate on net farm margin were investigated for suckler beef production systems and indicated that where N response is lower (i.e. in a later growing location) later turnout results in greater profitability. In other words, on farms where the grazing season begins later, turnout date (and calving date) should also be matched to this date to optimise profitability.

Currently it is estimated that, on average, grazed grass constitutes 49 per cent of the total feed budget on Irish suckler beef farms. Total herbage utilised is less than 5 t DM/ha. It is anticipated that the proportion of grazed grass in the annual feed budget could increase modestly and herbage utilised increase substantially (when stocking rate increases are also taken into consideration) by 2018.

The modest increase in grazed grass proportion in the annual feed budget is a reflection of a change in finishing systems from a grass-based extensive production system finishing steers at grass at 28 months to a more intensive system finishing steers indoors at 26 months of age. These targets are considerably lower than those set for the Derrypatrick Herd in Grange, where grazed grass and herbage utilised is estimated to account for 60 per cent of the total feed budget and over 10 t DM/ha, respectively.

Calving Date

Suckler beef production in Ireland is predominantly based on spring-calving cows with 70 per cent of calvings between January and May. However, there continues to be an interest in autumn-calving systems. A key motivation for autumn-calving in many cases is to provide weanlings for the premium priced, live export market.

This market requires E and U grade weanlings and in this respect, autumn-calving systems facilitate greater use of AI as cows are indoors during the breeding season, thus providing for increased sire selectivity and higher quality (muscularity and weight for age) progeny. Autumn-born weanlings are also available for sale earlier in the season and can therefore, avoid the peak weanling supply period in late autumn.

Where sale is delayed until this peak supply period, sale live weight is greater and hence, weanling/yearling value is also greater. Where a split-calving pattern is operated, i.e. calving a proportion of the cow-herd in spring and the remainder in autumn, a further advantage is that labour requirements are not concentrated into a single period. However, autumn-calving systems are associated with higher costs relative to spring-calving systems.

Firstly, feed costs are typically greater because the cow is lactating during the winter indoor feeding period and requires higher quality (more expensive) silage and/or concentrate supplementation.

Secondly, housing/facility costs are greater as additional creep areas for calves are required. The impact of calving season on net farm margin for suckler calf-to-weanling systems.

Three results are apparent: 1) Spring-calving systems are more profitable at all weanling prices. 2) The profitability of autumn-calving systems increase at a greater rate as weanling price increases. In essence, the additional weanling price is captured to a greater degree by the additional liveweight output from autumn-calving systems. 3)

A weanling price of €167/100 kg and €204/100 kg is required to breakeven in spring-and autumn-calving systems, respectively.

For spring-calving systems the date of calving is also of interest i.e. what is the optimumspring-calving date. If mean calving date is too early, i.e. prior to the start of the grazingseason, lactating suckler cows will require supplementary feeding and/or higher digestibility (more expensive) grass silage.

Conversely, if calving date is delayed until after the grazing season begins, the economic advantage of early spring grazing will not be captured i.e. dry, pregnant cows will remain indoors on more expensive grass silage despite the availability of cheaper grazed grass. Research at Grange has shown that delaying calving date by three weeks or six week reduced profitability by nine per cent and 19 per cent, respectively.

This equates to a reduction in profitability of €1.41/cow for each day that calving date is delayed. In this analysis, no performance effect of turnout date is assumed and therefore, the effects are due to differences in feed costs and slurry handling costs

Adding and Capturing Calf Value

Cattlemen may hear about the shrinking beef industry and wonder about their role in the future.

They can take heart in the expanding high-quality end of the business, however. Licensed partners of the Certified Angus Beef brand worldwide sell more than 2 million pounds of branded product per day, and supply has increased 92 per cent in the past five years. That's according to Mark McCully, CAB assistant vice president, supply.

Addressing producer-members of the Pittsylvania County Cattlemen's in Chatham, Virginia, earlier this month, Mr McCully said demand for CAB products grew along with supply, and that represents opportunity for producers.

Overall higher cattle prices and premiums for the best cattle are two results of strong demand, but producers can take steps to move a greater share of their calves into that premium category, he said. Genetic selection tools available on registered Angus cattle, specifically EPDs (expected progeny differences) can help any herd make progress.

"EPDs are used to compare animals within a breed but you should also pay attention to the average values of the breed," Mr McCully said. "For example, using a bull in the top half of the breed for Marbling EPD, or those above +0.40, is more in line with a genetic focus on the CAB brand."

Given the genetic potential, cattlemen can see it realized through comprehensive herd health and nutrition programmes and low-stress management, he added. "Then, find ways to get carcass data by working with organised state programmes, your bull supplier or a CAB licensed feeding partner. Keep detailed records and use that data in sire selection and cowherd culling."

To illustrate value differences, Mr McCully shared three scenarios with 750-pound (lb.) feedlot calves. Groups one and two were both age-and-source verified, gained 3.5 lb. per day (ADG) with feed-to-gain (F:G) conversion of 6.1. After a theoretical one per cent death loss, both groups finished at 1,325 lb.

The key difference was in carcase grading: Group 1 had five per cent Prime, 40 per cent CAB and 90 per cent Choice or better along with 30 per cent Yield Grade (YG) 1 or 2 and 15 per cent YG 4. Group 2 was leaner with 40 per cent YG 1 or 2 and just five per cent YG 4, but no Prime, only 10 per cent CAB and 50 per cent Choice with five per cent Standard. Then there was Group 3, the calves without age-and-source verification, ADG at 2.9 lb., F:G at 7.0, apparently in poorer health with 4 per cent death loss and finishing at 1,250 lb. They managed the same carcass results as Group 2, but came in $195 per head lower value than Group 1 under current market conditions.

Mr McCully concluded by emphasising the importance of marketing options to capture the value in "value-added" calves.

- Retain full or partial ownership of calves through the feedlot
- Direct marketing to feedlots with bonus options for carcass premiums
- Commingled sales of high-quality calves with other like-minded producers
- Calves backed by a resume that documents their profit potential
- Age-and-source verification with AngusSource®, which generally returns at least $25 per head.

Calving Pattern in Suckler Herds

It's important to have a tight calving pattern in most herds for a number of reasons, according to XL Vets.

A tight calving pattern will:

- Make management easier;
- Have all cows at the same stage of the breeding cycle;
- Make feed and grazing management easier;
- Allow batching of management tasks such as disbudding, castration, tagging, vaccinations, worming, pregnancy diagnosis etc.
- To optimise the timings of the above tasks;
- To improve management during calving;
- To give more even batches of calves for sale.

In order to have a tight calving pattern you have to want one and be prepared to work to achieve it. Everything has to be managed optimally to maintain a short calving but this is more than offset by the reduced work once it is in place.

Starting with your bulls; they need to be fit not fat so for spring calvers leaving the bulls out overwinter with a bit of mouldy hay will often leave them too little time to recover before their work begins. Turning a fat young bull out straight from the sales often results in a rapid loss of condition and the bull failing to work. Conformation needs to be good.

A bull with huge hind quarters that has weak hocks and can't serve his cows is no use to anyone. Over grown and infected feet need sorting out well before the bulling period as they can affect fertility. Get your Veterinary Surgeon to check the fertility of your bulls. This will pick up those producing poor semen and some physical defects but you still need to see the bulls working. A bulls ability to serve is affected by his enthusiasm or libido and a range of physical conditions such as corkscrew penis, deviated penis, adhesions, persistent frenulum, ruptured penis as well as lameness caused by infection, arthritis or osteochondritis.

Moving onto your cows again maintaining or improving condition is probably the most important factor in good fertility and a tight calving pattern.

Thin cows are unlikely to cycle and if they do they have less chance of holding in calf. Fat cows are more likely to have calving problems which will leave them less fertile. Nutrition is the main factor for your cows they need adequate energy to cycle.

A correct protein level to balance the energy is also important as are adequate mineral levels. Copper and selenium in particular are important in fertility though other trace elements such as iodine are relevant. Ease of calving plays a big factor in subsequent fertility and without high levels of fertility a tight calving pattern is not possible.

A cow that easily by herself produces a small to medium sized calf after a short gestation period will have a much better chance of being ready for the bull at the start of the mating period. There is much less chance of damage or infection when compared to a cow that holds onto her calf over 300 days and then struggles to calve. So breed selection and selection within a breed for desirable calving characteristics are very important if you want a tight calving pattern with little wastage.

There are infectious conditions to control to maintain optimum fertility. You can vaccinate to protect against Leptospirosis, Bovine Viral Diarrhoea (BVD), Infectious Bovine Rhinotracheitis (IBR) and in some circumstances Campylobacter or you may improve biosecurity to keep the diseases out once you know you have eliminated them.

The Scottish Agricultural College runs a Premium Cattle Health Scheme which monitors disease status and can include Johnes another disease that can reduce condition and fertility in infected animals. Parasites will also have an effect on condition and performance so fluke, worms and lice should be monitored and treated accordingly.

Maintaining a tight calving pattern can bring great benefits but to achieve one requires some effort. Effective health planning with your vet should make the task easier. It is possible to get good conception rates in a six or seven week bulling period if everything is performing to its optimum.

When to Introduce Heifers to the Milking Herd?

Introducing newly calved heifers into the main herd after the afternoon milking, reduces exposure to aggression and improves welfare, according to new research from the Agri-Food and Biosciences Institute. This is just one of the interesting findings of a study, carried out by researchers in Northern Ireland, the results of which were presented to delegates at this year's British Society of Animal Science's annual conference, held at Queen's University, Belfast.

The integration of dairy heifers into the main herd during the post calving period can have negative effects on the heifer's welfare and productivity and this appears to be related to the fact that heifers attain low social status when entering the milking herd and therefore are subjected to increased levels of bullying and aggression.

The aim of this study was to assess if the time of day, in other words after morning or afternoon milking, when freshly-calved heifers are introduced to a group containing mature cows influences the heifers' welfare and performance," said the Agri-Food and Biosciences Institute's Alastair Boyle, explaining the rationale behind his team's work. A total of 28 Holstein Friesian heifers were used in the study, with heifers introduced into a group containing mature cows between approximately 24 and 36 hours post calving. Animals were housed in cubicle pens with solid floors. Two treatments were examined, with heifers introduced either after morning milking or after evening milking. Treatments were replicated five times, with treatments balanced for genetic merit, body weight, condition score and sire. In each replicate a resident group, containing 12 cows and six non-experimental heifers, was established at least seven days prior to the introduction of the first experimental heifer.

The social and exploratory behaviour of each heifer was recorded directly during a two-hour period immediately after introduction to the group on day one. These behaviours were observed for each experimental heifer during four five-minute continuous observations at 30-minute intervals during the two-hour post-feeding period on one day each week for one month post introduction.

"And we saw that on the first day animals in the 'morning' treatment received more threats and butts than those in the 'afternoon' treatment," said Mr Boyle.

"There was a tendency for heifers introduced at in the morning to be chased more than those introduced in the afternoon and heifers introduced to the group in the afternoon were more socially cohesive than those in the morning treatment.

"No significant treatment effect was found for total hours lying, with heifers in both treatments lying for an averaged 7.5 hours/day. During the first day heifers in both treatments lay for fewer than 4 hours/day," he added.

Bovine EID Update

At the end of February, the European Commission announced that it was working on a proposal to introduce electronic identification in bovine animals. TheCattleSite Editor Charlotte Johnston speaks with Sergio Pavon, the European Commission official in charge of animal identification and traceability to find out more.

At the Paris 'Salon de l'Agriculture' (19th-27th February) the European Commission announced it is preparing a proposal for electronic identification in bovines, upgrading the existing plastic tagging system.

DG SANCO, the European Commission's Directorate General for Health and Consumers, is confident this step will further increase the traceability of animals and food products for veterinary and food safety purposes, controlling the spread of animal diseases, with veterinarians playing a key role in the process.

1) Would bovine EID be compulsory and how much would this cost producers? It is uncertain for the moment whether bovine EID would be implemented on a compulsory or on a voluntary basis.

2) What maintenance costs would be involved? The cost of EID tagging is more expensive than the cost of conventional ID tagging. This is a general rule that applies to all Member States (countries in the EU). However, the price of the identification devices will largely depend on how the ordering of the tags is organised within an EU member state. If the tags are tendered on a regional, or on a national level, or if every independent farmer has to order his or her own tags, the price of the tag sets could increase by up to 40 per cent.

3) Would grants be offered to help roll this out? EU legislation does not determine who should cover the costs of EID. Member States may financially support farmers for the introduction of EID within the framework of EU rules on state aids. Support is possible under different Commissions' rural development measures.

4) What are the benefits for producers? E.g. would they store records of vaccinations, birth weights etc. EID, if in conjunction with e-reading, may help to reduce identification errors and reduce time to search the history of each bovine. One of the main benefits of EID if used in conjunction with e-reading and e-transfer, would be the reduction of administrative work, such as written notifications about the bovines' identities, for keepers (farmers and other stakeholders).

Currently, all bovine notifications (births, deaths, animal movements) must be manually registered and converted into an electronic format to the computerised database. EID will also support the competitiveness of the sector as an excellent tool for improving farm management and on-farm automation.

Additional benefits of bovine electronic identification cover disease prevention and control, preventing fraud, genetic improvement, crisis management and trade will be among them. In addition to strengthening the current system of traceability, consumer protection and food safety. Food processing establishments (e.g.slaughterhouses) and markets will benefit, in addition there will be a reduction of the administrative work and labour costs.

Improve Calving Weight

The breeding programme; which bull was used to breed a cow or group of cows, determines the genetic potential of the newborn calf or group of calves," says Barry Yaremcio, beef/forage specialist with Alberta Agriculture and Rural Development. "Birth weight, frame size and growth rate are all genetically controlled and cannot be changed. On the other hand, management on farm or on the ranch impacts the animals' growth potential. You can affect the size and weight of the calf come weaning time and, therefore, your profitability. If poor decisions are made, the genetic potential of the calves will be negatively affected.

"Management changes must be evaluated, not just made on a whim. The change or changes should move the operation towards long term goals and make economic sense. Be it a change in bull genetics, or changes in the feeding and grazing programmes, the short and long term benefits need to be greater than the added costs or potential down side."

There are several things producers can do to improve calf weaning weights without wholesale changes:

Watch body condition score of the cows prior to calving-colostrum quantity and quality are impacted by body condition of the cow. If a cow is thin, less colostrum is produced, reducing the passive immunity provided to the calf. This can lead to a greater risk of disease or infection in these calves.

A cow in good condition is able to mobilize fat reserves. A pound of fat provides enough energy to produce seven pounds of milk. This is very important for the first 12 weeks of lactation when feed intake is limited. Castrate or ring bull calves at birth – the practice of leaving bull calves intact until weaning to improve growth rates has been disproven. Bull calves do not reach puberty until 9-to 10-months-of-age. At puberty, bull calves start producing hormones to increase growth rates. Unfortunately, most calves are castrated before this time. As bull calves grow, the stress associated with castration increases. An 8-or 9-month-old calf may experience reduced average daily gain of 0.3 to 1.3 pounds per day throughout the 30-day period after castration, adding up to a total 'lost' gain of 39 pounds.

Move cow calf pairs to a clean open area – keeping newborn calves in a tight pen with many other animals presents ideal conditions for diseases to spread. If possible, move cow-calf pairs to clean pens away from cows that have not calved, or to pastures. Minimizing calf contact with other animals, and standing water puddles contaminated with urine and manure, reduces disease pressure.

Use a fortified trace mineral salt with selenium – in most parts of Alberta, copper, manganese, zinc and selenium are deficient. Cobalt and iodine are virtually absent from all forges and grains. The use of a cobalt (red) or cobalt/iodized (blue) salt does not provide any of the four trace minerals mentioned above.

Changing from a blue salt on pasture to a fortified trace mineral salt with selenium resulted in an additional 20 pounds of weaning weight. There can also be improvements in cow fertility when the trace mineral salt with selenium is fed year round.

Develop a water system where cows do not walk into the water supply – water quality impacts average daily gain. Work done in the Barrhead/Westlock area in the early 1990s, found an improvement in calf growth rate if stock drank from a nose pump or from a storage tank. Over the grazing season, there was a 20 pound improvement in calf weight gain.

Creep feed calves – cow milk production peaks about eight weeks after calving. After this, milk yield declines. Creep feed should be provided when nutrients from the fresh grass and milk do not provide what the calf requires. Grazing management, weather, moisture and pasture fertility all impact when the creep feed should be provided.

Calves that consume creep feed prior to weaning have less difficulty adjusting to a ration that includes grain after weaning. Manage forage resources – let the grass grow in the spring before turning the cows out to graze. There should be a minimum of three to four leaves of new growth to allow the plant to start replenishing food reserves in the root zone. For continuous grazing systems, every day the cows are turned out too soon in the spring, will result in a reduction of fall grazing of two to three days if a continuous grazing system is used.

Forage quality drops rapidly after the seed head has emerged from the boot. The use of rotational grazing to allow plants to recover after grazing events and to keep grass vegetative, maintains quality and increases potential yield. When a seed head is fertilized, the plant allocates nutrients to fill the head rather than initiating vegetative growth; overall yield potential is reduced.

Minimize weaning stress – vaccinating calves four to six weeks prior to weaning provides a higher level of immunity compared to calves vaccinated at weaning. Having a stronger immune system helps combat disease and the associated reduction in animal performance. "Prior to weaning, move the calves and cows into the pens or corrals where the weaned calves will stay after weaning," says Mr Yaremcio. "Let the cows teach the calves where the feed bunks, feeders, water bowl and loafing areas are. When the cows are moved out of the pen, the calves know where everything is. Less stress results in less illness and reduced performance."

Once weaned, place the cows in an adjacent pen to the weaned calves. Allowing the two groups to see each other and touch noses also reduces stress. "Even if you are selling the calves at weaning, inform the buyers that the calves have been vaccinated," says Mr Yaremcio. "This information may entice the buyers to add a few cents a pound to the settlement price."

Calf Health

The aim of successful calf rearing is to produce a healthy calf which is capable of optimum performance throughout its life from birth through to finishing, reports Bernadette Earley, Animal & Grassland Research and Innovation Centre, Teagasc.

The aim of successful calf rearing is to produce a healthy calf which is capable of optimum performance throughout its life from birth through to finishing. A suitable calf rearing system has the following characteristics:

- Good animal performance with minimal disease and morbidity
- Low cost input

- Low labour input.

To ensure a healthy calf, the aim is to minimise the calf's exposure to disease, and maximise its defence against disease. In minimising a calf's exposure to disease, providing a clean, disease-free environment is fundamental. This involves:

- Thorough cleaning and disinfection, before and during the calving season, of all areas used by calves.
- Providing a clean, straw-bedded lying area with no draughts and good ventilation.
- Accommodating calves in batches so that young calves are never mixed with or accommodated in areas used by older calves.

The objective of a well-designed herd health programme is to address multiple areas of management in order to reduce the likelihood of disease outbreaks and is a necessary step if economic returns are to be realised. A herd health programme that includes bio-security, vaccination and the culling of carrier animals, drawn up in consultation with a veterinary practitioner, is the best way to address disease problems.

Infectious Disease Agents

Calf Scours

Scours are the main causes of calf mortality. The majority of calf scours are caused by six organisms: viruses such as rotavirus and coronavirus, bacteria such as E. coli and Salmonella sp., and protozoa, such as cryptosporidia and coccidia. As outlined above, vaccination of the dam will help reduce the probability of calf scours but cannot solely be depended upon for prevention.

Furthermore, there are no vaccines available to combat protozoa. However, good hygiene and management practices will reduce the likelihood of infection from cryptosporidia and coccidia. Diarrhoea in calves results in losses of water and electrolytes, such as sodium, bicarbonate, chlorine, and potassium. Calves with diarrhoea can lose 10-12 per cent of their body weight in water losses. Depending on the severity of the diarrhoea and dehydration, calves may need to receive oral electrolyte solutions once daily or as many as four times a day. Calves should be fed their regular allowance of milk when receiving oral electrolytes. Scientific evidence has accumulated that continued milk feeding does not worsen or prolong the course of diarrhoea, despite a somewhat lowered digestive capacity. The milk supplies the calf with energy and other nutrients that are essential for survival.

Calf Pneumonia

The underlying cause of pneumonia or bovine respiratory disease (BRD) is extremely complex with the involvement of viruses, bacteria and mycoplasma. The incidence of infection is usually high, but the mortality rate is variable. The main viruses that cause outbreaks of pneumonia in calves are IBR, RSV, PI-3, and BVD. Factors associated with susceptibility to pneumonia in calves are; stress (disbudding, castration), overcrowding, inadequate ventilation, draughts, fluctuating temperatures, poor nutrition and/or concurrent disease. In most cases the main infective agent is a virus, which causes respiratory tract damage.

This effect is worsened by Mycoplasmas and secondary bacterial infections (e.g. Mannheimia (Pasteurella) haemolytica). Viruses are unaffected by antibiotics, however, antibiotic treatment is usually administered to kill off the secondary bacterial infections and offer the calf the opportunity to fight the disease. In order to direct the appropriate treatment strategy, nasal swabs should be submitted to the Regional Veterinary Laboratory for accurate identification of the pathogen(s) involved. Calves should be vaccinated where specific problems arise. Veterinary advice should be sought and the widest protection against pneumonia will be achieved where a vaccination programme includes the three most common respiratory viruses (IBR, RSV and PI-3) and the bacterial pathogen Mannheimia (Pasteurella) haemolytica.

Calf Immunity

In the bovine species, immunoglobulins do not cross the placenta in utero, and the newborn calf is, therefore, dependent on antibodies obtained through ingestion of colostrum. Maternal colostrum provides the main source of immunoglobulins (Ig) and other nutrients for the newborn calf. If the serum immunoglobulin concentration is less than 10 mg/ml when sampled between 24 and 48 h of age, calves may be defined as having failure of passive transfer (FPT) of protective colostral immunoglobulins. Calves that receive inadequate colostrum are more susceptible to neonatal infections. This problem can be particularly severe in calves that have been moved off their farm of origin and through markets. In these circumstances, there is greater risk of exposure to infection. Foetal growth retardation or a stressful birth process is likely to reduce the ability of the intestine to absorb immunoglobulins from colostrum.

Variation in Immunoglobulin Concentration

Immunoglobulins help to maintain the animal's health and reduce mortality rates by helping to eliminate foreign agents in the body (e.g. bacteria and viruses). Considerable variation exists between cows with respect to immunoglobulin concentration in the colostrum. No significant difference between colostrum IgG1 concentrations in either the front or hindquarters of the udder have been reported however, a large variation in colostrum yield exists between beef and dairy cows. Research at Teagasc, Grange has clearly shown that dairy calves with low levels of immunoglobulins had the highest incidence of diarrhoea, respiratory disease and mortality compared with age matched suckled beef calves (Earley et al., 2000). This is primarily due to the much lower concentration of IgG in colostrum of dairy cows compared with suckler cows.

Factors Influencing Immunoglobulin Concentrations in Calf Serum

The main factors influencing immunoglobulin concentrations in calf serum are:

1. Time of feeding/suckling,
2. Volume of ingested colostrum and
3. Immunoglobulin concentration in colostrum. Consequently, with regard to colostrum feeding, there are a number of main points which must be noted.

Time of Feeding

- The immunoglobulins in colostrum must get into the calf's blood via absorption from the small intestine.
- The ability of the calf to absorb these immunoglobulins decreases linearly after birth and generally stops by 24 hours of life.

This means that the earlier a calf is fed/suckles after birth the greater the level of immunoglobulin absorption. Ideally, the calf should ingest colostrum within one hour of birth. There are also advantages in continual colostrum ingestion after the first day as the immunoglobulins in colostrum also acts locally in the gut and helps fight against septicaemia.

Colostrum-Feeding Procedures

It is well recognised that the oesophageal groove reflex is not triggered when colostrum is administered by the oesophageal tube method and this results in colostrum deposition in the forestomachs. In contrast, when a nipple bottle or bucket is used, suckling triggers the oesophageal groove reflex, resulting in the deposition of colostrum directly into the omasum and abomasum, where it can quickly empty into the small intestine to be absorbed. The ability of the intestines to absorb IgG starts to decline progressively after 4 to 6 hours and ceases after 24 hours from birth. This means that the earlier a calf is fed/suckles after birth the greater the level of immunoglobulin absorption.

Current recommendations for normal sized dairy calves are to administer either 3 L of good quality colostrum within 1-2 hours of birth by oesophageal tube (Chigerwe et al., 2008) or to ensure that calves receive at least 2 L within 4 hours by nipple feeding and a total of 4 L within 12 hours from birth (Chigerwe et al., 2009)

Research Studies – Calf Health

In a Teagasc study, 93 Charolais×Friesian (Ch×Fr), 30 Limousin×Friesian (Li×Fr) and 100 Holstein-Friesian (HF), calves were purchased directly from marts and were approximately 21 days of age on arrival at Teagasc, Grange Research Centre. Calves with a rectal temperature greater than or equal to 40°C and clinical signs of pneumonia were administered antibiotic for the treatment of clinical symptoms (defined individually for each animal). Fifty-four out of a total of the 223 purchased calves remained healthy throughout the 63 day rearing period indoors. The incidence of respiratory disease was higher in calves with low serum Ig levels and ZST units. Calves with low immunoglobulins (less than 10 ZST Units; less than 21mg/ml total serum Ig) are more susceptible to respiratory disease.

In another Teagasc study, Earley et al. (2000) quantified serum immunoglobulin concentrations in suckled beef calves and dairy calves. The spring-calving suckler herd at Grange was the source of suckled calves.

This herd, which calved indoors, consisted of the following breeds: Charolais, Simmental×(Limousin×Friesian) and Limousin×Friesian. The calves were either by a Charolais or Limousin bull and consisted of: Continental types (representing the progeny of Charolais cows by either Charolais (n = 8) or Limousin bulls (n=8)), ^! Continental × [!Friesian (representing the progeny of the Simmental × (Limousin× Friesian) cows and Charolais bull) and ¾Continental×¼ Friesian (representing the progeny of the Limousin×Friesian cows and Charolais (n=52) or Limousin (n=12) bulls). Calves were then left with their dam and blood sampled 28 and 56 days post partum. Male spring-born calves from dairy herds were purchased from auction marts and consisted of Charolais×Friesian (n=61), Limousin×Friesian (n=39), Belgian Blue×Friesian (n=9), and Holstein-Friesian (n=73). The dairy calves were approximately 28 days of age on arrival at Teagasc, Grange Research Centre.

They received an individual allowance of 25 kg of milk-replacer powder, offered warm (38oC) by bucket (125 g/1), during the first 42 days and had ad libitum access to a concentrate ration during the first 56 days after arrival. Serum samples were collected from the dairy-herd calves on days 0 (day of arrival) and again 28 days later. Serum samples were collected from the suckled calves 28 and 56 days post partum. The total Ig serum concentrations and ZST units were significantly higher in the suckled calves compared with the purchased calves at 28 and 56 days of age. In the present study, all suckled calves were individually fed 40 ml of first milking colostrum per kg of birth weight from the calf's own dam, within one 1 h of birth, using an oesophageal feeder and bag (stomach tube). The purpose of this procedure was to ensure suckled calves received an adequate quantity of Ig of known quality (> 160 mg IgG/ml) in the immediate post-partum period. The colostrum feeding procedure for the dairy-herd calves used in the present study was unknown. The marked differences in total Ig concentrations between suckled calves and purchased calves suggest that calves born in dairy herds received either insufficient quality or quantity of colostral immunoglobulins.

Research Studies-Requirements of the Housed Calf

The reasons for housing artificially reared calves are mainly management ones. The calf born outdoors is capable of finding its own shelter. In the confinement of a house, away from its mother, the calf needs to be provided with:

- A dry surface to lie on
- Protection from draughts
- Adequate ventilation.

If these criteria are met the healthy calf should be insensitive to weather changes outside the house, i.e. temperature, humidity and wind speed. Dry, draught free housing will reduce the environmental stresses on calves and adequate air changes resulting from good ventilation reduce the infection load on the calves. It may not prevent pneumonia. However, the severity of pneumonia will be less and the mortality associated with it will be reduced. From an animal health and welfare viewpoint it is important to develop a combination of management procedures which will minimise the adverse effects of respiratory disease on calf performance and health/welfare indicators.

The effects of rearing calves outdoors using calf jackets on performance and immune responses with a view to developing management procedures to improve the health and welfare of calves was investigated in a study at Grange. The study showed that rearing calves outdoors using calf jackets had no beneficial effect on calf performance. The incidence of respiratory disease was higher in calves reared indoors when compared with calves reared outdoors with and without jackets. There was an increased incidence of diarrhoea in calves reared outdoors irrespective of calf jacket (Earley et al., 2004).

The choice of calf house will depend to a large extent on the size of the enterprise and on labour availability, feeding system, penning arrangement and intensity of use. Ideally the calf house should meet the environmental requirements of the calf and accommodate the feeding routine of the rearer.

A number of studies at Teagasc, Grange have shown that there was no significant difference in calf performance between the naturally ventilated calf house and the calf hutches. While there was a significant reduction in the number of calves treated for respiratory disease among calves reared in hutches (58% versus 43%), it was evident that the isolation provided by the hutches did not markedly reduce the incidence of respiratory disease among bought-in calves which may already have been exposed to infectious organisms (Fallon, 2009).

A series of four experiments used 320 calves to compare a conventional naturally-ventilated calf house with a similar sized building where ventilation was provided by a fan system which forced air into the building. On arrival at Grange the calves were allocated to the following treatments:

1. Climatic naturally ventilated calf house
2. Fan ventilated-forced air, calf house.

There was no difference between houses in the incidence of disease, mortality, feed intake or liveweight gain in any of the four studies. It was concluded that the fan system which forced air into the building gave similar calf performance to a naturally ventilated building irrespective of disease challenge.

Calves are exposed to infectious organisms from the moment of birth, and natural defence mechanisms usually prevent the establishment of disease. Animals develop disease because of a complex relationship between the host (animal), the infectious agent (bacterium, virus etc.), and the environment. Control of the agent is largely based on prevention of exposure, immunity, and chemotherapeutic (drugs) agents. In maximising a calf's defence against disease, control measures include:

- Adequate nutrition of the pregnant cow.
- Vaccination of cows for control of any organism(s) known to be responsible for infection on the farm in calves e.g. E. coli, Salmonella, rotavirus and coronavirus. In this respect, vaccination alone is not a replacement of good management, good hygiene or good biosecurity.
- Disinfecting the calf's navel immediately after birth.
- Pooling of colostrum is discouraged for biosecurity reasons.
- Ensuring that each calf receives sufficient colostrum (first milk) immediately after calving. Colostrum provides not only food but also maternal antibodies (immunoglobulins) to protect the young calf against the common infections that it is likely to encounter in early life.
- Regular temperature checking is useful to guide both diagnosis and observation of a clinical problem.
- A veterinary practitioner should always be consulted with regard to specific health problems.

High Concentrate Diets For Dairy Bulls

Finishing bulls on high concentrate diets is a high cost system with a large input of concentrate feeds. For the system to be profitable concentrates must be cheap and calves must be relatively cheap and of high growth potential. This system requires careful diet formulation and excellent feeding management to ensure maximum performance and minimum risk of digestive upsets. This paper will outline target intake levels as well as guidelines on diet formulation and appropriate feeding management.

Target Weight Gain

At best, this is a guideline to target gains as actual LWG will be dictated by a number of factors including the previous history of the animal, weight for age, diet type, feeding management, husbandry etc.

The scope for compensatory growth depends on the previous nutritional history of the animal. Lifetime performance to the start of the finishing period will have a major bearing on animal performance over the finishing period. The lighter an animal's weight for age, the greater the animal's potential for compensation. Differences in compensatory growth potential can result in big variations in response to concentrates. Animals with little compensatory growth tend to have poorer performance and efficiency.

Duration of the Finishing Period

Unlike steers, where performance starts to decline after 80-90 days on *ad-libitum* meals, young bulls have been successfully fed for up to 270 days, but rate of gain declines significantly over time. The duration of the meal feeding period will be dictated by the start weight, carcass weight required, level of fatness and other factors such as lameness, particularly for bulls on slats. It will also depend on where an animal is on the growth curve. Animals with a lot of growth potential might be expected to continue thriving for longer than an animal with high lifetime performance to that point. As a rule of thumb a maximum feeding period of 170-180 days is preferable. Starting weight for high concentrate feeding of dairy bulls will be dictated by target slaughter weight but assuming a liveweight gain of 1.25 kg/day, animals should be within 220 kg of slaughter, when high concentrate feeding begins. High ratesof gain over short periods of time are very efficient, all else being equal.

Dry Matter Intake

Achieving high dry matter intake is essential for good performance. The level of intake achieved by dairy bulls in research studies at Teagasc, Grange. Young calves at 12 weeks (Batch 1) on an *ad libitum* concentrate diets achieved an intake of 2.8 per cent of body weight but this declined significantly over the next 140 days. The overall intake from 110 kg to 442 kg body weight (Batch 2) was 2.2 per cent of body weight. On average, heavier animals might be expected to achieve 2.0-2.1 per cent of body weight on an *ad libitum* feeding system. Experience would suggest that dairy bulls will achieve intakes of 2.1-2.2 per cent of BW for a high concentrate feeding period of 170-180 days but this will decline over time to 1.8-1.9 per cent of BW, with slaughter weights of 500-600kg.

Feed Efficiency

The feed conversion efficiency achieved on high concentrate diets with dairy bulls across six experiments is present. The average feed conversion efficiency was 5 kg DM feed per kg liveweight gain for animals with an initial weight of 98 – 230 kg and a final weight of 448-454 kg. Similar studies have recorded feed efficiency of 9.0 kg DMI per kg carcass gain for similar animals.

Considerable work has been done at Teagasc Grange comparing dairy bulls slaughtered at different slaughter weights. Animals were fed high concentrate diets for six months (179 days) or nine months (272 days). Daily gain over the first six months was 1.4 kg/day. This fell to 1.2 kg/day for the period from six months to nine months. Slaughter weight was 550 kg after six months and 670 kg after nine months. Total concentrate consumption was 1.76 t for six months and 3.0 t for nine months. Feed conversion efficiency was 7.2 and 8.1 kg DMI/kg LWG for the bulls slaughtered at six and nine months, respectively. Considerable work has been done at Teagasc Grange comparing dairy bulls slaughtered at different slaughter weights. Animals were fed high concentrate diets for 6 months (179 days) or 9 months (272 days). Daily gain over the first 6 months was 1.4 kg/day. This fell to 1.2 kg/day for the period from 6 months to 9 months. Slaughter weight was 550 kg after 6 months and 670 kg after 9 months. Total concentrate consumption was 1.76 t for 6 months and 3.0 t for 9 months. Feed conversion efficiency was 7.2 and 8.1 kg DMI/kg LWG for the bulls slaughtered at 6 and 9 months, respectively.

Diet Specification

Energy, protein, fibre and minerals are the primary components of the diet that need to be balanced correctly.

Energy

High energy feeds should be fed for maximum weight gains. Dietary energy density will depend on concentrate cost, the response in carcass gain from increasing/decreasing energy density and the corresponding value of carcass gain. The minimum energy density in the concentrate mix should be 0.95 UFV/kg as fed or 1.09 UFV/kg DM. Grange studies, with young bulls slaughtered at 12 months, showed that when a low energy concentrate feed was used, compared to a high energy feed, carcass weights were up to 28DM. Grange studies, with young bulls slaughtered at 12 months, showed that when a low energy concentrate feed was used, compared to a high energy feed, carcass weights were up to 28 kg lighter on the low energy feed.

The selection of feed ingredients for this system is critical. The primary energy sources are based on starch or digestible fibre. Cereals such as barley, wheat and maize grain are based on starch, while citrus pulp, beet pulp and soya hulls are based on digestible fibre. A mixture of these energy sources is preferable to stimulate intake and reduce the risk of digestive upsets such as acidosis.

The mix being used at Johnstown Castle and the nutritional value (on a fresh weight basis) is presented below.

One source of energy e.g. cereal may be used in the high concentrate system, provided it is correctly balanced for protein, minerals and long fibre. It is imperative that feeding management is excellent. This system is inherently more difficult to manage and there is a greater risk of digestive upsets.

Protein

During the growing phase bulls need adequate protein to build a frame and to ensure that they do not remain small and fat. Finishing animals will have a lower requirement for protein because their frame is already established and high levels of protein are not required to lay down lean & fat tissue. Protein may be more critical with Holstein dairy animals that tend to continue to grow rather than lay down tissue. The crude protein content of the concentrate will depend on the composition of the diet, for example where sugar beet, fodder beet or other low protein feeds are being offered a higher crude protein in the concentrate will be necessary.

Roughage Source

Roughage is required for the satisfactory functioning of the digestive system. At least 10 per cent of the DM intake must be a source of long roughage but this should not exceed 15 per cent. This may be silage, straw or hay. Anecdotal evidence would suggest that straw/hay is the preferred option. If feeding hay it is important to ensure intake is not excessive as this will affect performance, but otherwise it is a safe feed. Roughage may be fed separate or mixed with the concentrate. If feeding separately, it is important to ensure animals have adequate access to the roughage and it is regularly freshened up.

Minerals

Cattle need minerals to maintain good health. This is particularly important on a high concentrate feeding system where issues such as hoof health may be compromised if animals are on a high concentrate diet for long periods of time. It is important that a standard beef ration is not used for ad libitum feeding as the mineral specification will be in excess of requirements and may lead to poor performance and the risk of toxicity and severe diarrhea.

Grain Processing

Ground cereals should not be used in high concentrate feeding systems. Pelleted rations are generally manufactured with ground cereals – check with the manufacturer.

Water

Water intake will be high on this system. Animals on ad libitum concentrate have a high rate of metabolism and dissipate a lot of sweat. Fresh water must be available at all times. It is recommended that water troughs be inspected daily and cleaned at least 2-3 times a week or immediately if water is fouled. Animals must never be left without water on this system. Inadequate water intake will depress feed intake and consequently performance.

Feeding Management

Feeding management is critical. There are a few basic rules that must be adhered to:

Introducing the Concentrate

An introductory period of approximatnely 3 weeks is necessary to allow animals adapt to a concentrate based diet. Start off on 3 kg concentrates and every 4 days add 1.5 kg concentrates. Animals should have full access to silage/roughage while being built up to high concentrate diets.

By gradually introducing cattle to a high concentrate diet, the risk of acidosis or feed sickness is reduced. At any sign of excessive scouring or digestive upset, decrease the concentrate level to the previous step for a few days. Animals should be watched closely and regularly during the introductory period. A culling policy should be put in place whereby problem animals are removed. These animals will most likely not perform throughout the finishing period, incur large costs and therefore result in reduced returns.

Frequency of Feeding

Subsequent to the introductory period once daily feeding should be satisfactory after the introductory period, if troughs are large enough. It is not safe to ever go below 5 per cent remaining in the trough. Do not have cattle waiting on empty troughs for the next feed.

Feed Trough Management

Good feed trough management is essential in preventing digestive upsets. It is essential to recognize that even slight digestive upsets will affect growth rates and therefore your profitability long before an animal is recognisably sick. Bird/vermin infestation can be a problem where animals are being fed high levels of concentrates. Methods to overcome this include the use of shorter troughs or preferably hoppers.

Housing

Providing less than recommended space will cause stress in animals, depress feed intake and consequently reduced animal performance. In general, slatted pens are not ideal for high concentrate feeding for long periods of time. Straw bedding or out-wintering pads are preferable options.

It is critical that the housing is thoroughly checked out before housing animals. The principles of good ventilation must be followed. These include adequate pitch of the roof (15o), as well as adequate inlets and outlets and space sheeting on the roof. The importance of air movement cannot be over-emphasised.

Animals on high concentrate feeding systems dissipate a lot of sweat to keep themselves cool. Signs of animals sweating excessively translates into lack of air movement in the house. This is energy wasted that could otherwise be used for liveweight gain. Feed intake and performance will suffer severely/unless this is dealt with.

Body Condition Scoring for Beef Cows

Beef cattle producers know that a cow's body condition at the time of calving will affect the health of her calf and her ability to breed back in a timely manner.

A Kansas State University animal scientist encourages producers to use a body condition score (BCS) system to determine their cows' condition.

"A body condition score on a beef cow is the closest thing we have to determining her nutritional status at a glance," said Chris Reinhardt, livestock specialist with K-State Research and Extension. "But scoring cows properly and really benefitting from this tool requires more effort and observation than simply looking and thinking, 'they look a little thin.'"

The reason for talking about BCS now is that there is still time to adjust the nutrient supply to get the cows into the target BCS by calving time, Mr Reinhardt said.

To evaluate an individual cow, he said, look at her topline, brisket, ribs, flank, round, and tail head. A borderline thin cow (BCS = 4) will clearly show three to four ribs first thing in the morning, will have no obvious fat depots in the brisket or tailhead, and the individual vertebrae along the topline will be visible. The cow still shows some muscle through the round, and she may appear "healthy but thin."

In a borderline fleshy cow (BCS = 6) the ribs and vertebrae will not be obvious, as they are covered by fat. The muscling through the round will be plump and full, but muscle definition is still apparent, and there will be small but noticeable fat deposits behind the shoulder, in the flank, brisket, and around the tail head.

The "ideal" or "target" BCS for cows at the time of calving is the BCS = 5, Mr Reinhardt said. This cow will show the last one to two ribs first thing in the morning before feeding and have good fullness of muscle in the round with definite muscle definition. In addition, the spine will be apparent, but individual vertebrae will not be discernable, and with no obvious fat deposits behind the shoulder or around the tail head.

"We would say this cow has a good bloom," he added.

A change in BCS (from BCS 4 to 5, for example) requires the addition of 75 to 100 pounds of live body weight, depending on the mature size or frame size of the cows, the livestock specialist said.

"If you're two months from the start of calving and would like to add 0.5 to 1.0 BCS, you'll need to feed the cows for maintenance, the last one-third of gestation, and an additional 1.0 to 1.5 pounds per day of gain. This means increasing the amount of good quality hay as well as the amount of supplement."

Mr Reinhardt said that thin cows (BCS 4 or below) can be separated off and fed at a higher plane of nutrition.

"The argument can be made that this creates 'welfare cows.' However, good record-keeping will indicate whether these cows are perennial 'hard-keepers' or if they are simply too young or too old to compete with the mature cows," he said. "If they're too young, another year of maturity should cure this; if they're too old, you may consider culling them after weaning time. The key here is that good record keeping allows you to cull intentionally based on productivity, not based on lack of observation and management."

Cows at a BCS 5 at the time of calving should provide adequate colostrum and nutrition for their calf and breed back in a timely fashion, Mr Reinhardt said. Cows that calve below a BCS 5 will delay their return to estrus and breed back late. If these cows do not maintain a 365-day calving cycle, they could, after one to two late breedings, effectively cull themselves due to being open at pregnancy checking time.

"Young cows are especially susceptible to this because they are gestating a calf, nursing a calf and still growing frame and muscle themselves," he said. "Unfortunately, young cows are the future of your herd and possibly your most progressive genetics. Hopefully these cows aren't culled simply for lack of nutrients."

Mr Reinhardt encourages producers to take time to critically evaluate the nutrient status of their cow herds this winter, and to use the body scoring system to manage the fertility and health of their herds going into next spring. "That way, you give yourself full control over the genetics of your herd for years to come," he said.

Developing High Quality Replacement Heifer

Energy balance or plane of nutrition influences reproductive performance in heifers and cows. Numerous studies have reported inverse correlations between postweaning growth rate and age at puberty and pregnancy rates in heifers were shown to be dependent upon the number displaying estrus prior to or early in the breeding season. Thus, rate of postweaning growth was determined to be an important factor affecting age of puberty, which influenced pregnancy rates. This and other research conducted during the late 1960s through the early 1980s indicated puberty occurs at a genetically predetermined size, and only when heifers reach their target weight can high pregnancy rates be obtained. Guidelines were established indicating replacement heifers should achieve 60 to 65 per cent of their expected mature body weight by breeding.

Traditional approaches for postweaning development of replacement heifers used during the last several decades have primarily focused on feeding heifers to achieve or exceed an appropriate target weight, and thereby maximise heifer pregnancy rates. Substantial changes in cattle genetics and the economy have occurred over this time, indicating traditional approaches should be re-evaluated. Intensive heifer development systems may maximise pregnancy rates, but not necessarily optimise profit or sustainability.

These systems require significant use of fossil fuels, cereal grains, and high capital investment in equipment and facilities. Cereal grains, often used as an energy source in heifer diets, detract from the system's sustainability due to growing demand for human food and ethanol production.

Furthermore, almost all studies on heifer development conducted over the last half century have focused on production to first calving with little information concerning effects of heifer development systems on lifetime productivity.

Since inception of target weight guidelines, subsequent research demonstrated the pattern of growth heifers experience prior to achieving a critical target weight could be varied. Altering rate and timing of gain can result in periods of compensatory growth and/or allow producers to limit supplementation to critical periods of heifer development thereby providing an opportunity to decrease feed costs.

For example, delaying heifer gain until 47 or 56 d prior to the breeding season did not negatively influence reproductive performance, but reduced the amount of feed needed (Lynch et al., 1997). In one year of this study, puberty was delayed in heifers fed to achieve lower early gains, but first-service conception rate tended to be improved in these same heifers.

Similarly, Freetly et al. (2001) found delaying gain until the later part of the postweaning period reduced total energy intake, but calving rate, age at calving, postpartum interval, and second year pregnancy rate were not impacted. These studies indicate that total energy intake, and possibly heifer development costs, may be reduced by limiting heifer gain early in the postweaning period followed by accelerated gains before the breeding season.

Review of Target Weight

As indicated previously, substantial research contributed to the guidelines of developing heifers to 60 to 65 per cent of mature body weight at time of breeding. Studies evaluating different postweaning rates of gain or target weights have used either different amounts of feed, or different types of feeds varying in energy and/or protein content to obtain differences in rates of growth.

A review of these studies conducted over the last several decades along with new research discussed later, indicates the association among BW, puberty and heifer pregnancy rate appear to have changed over time. Research reports published through the late 1980s have shown much greater negative effects of limited postweaning growth on age of puberty and subsequent pregnancy, where as more recent studies indicate less of a negative impact of delayed puberty on pregnancy response.

Several factors likely contribute to this change over time. Initial research corresponds to the industry shift from calving heifers at 3 years of age to calving at 2 years of age. Thus, selection pressure for age of puberty was probably minimal in animals in the early studies. While selection intensity would have increased with the reduction in calving age of heifers, genetic progress would take time due to the long generation interval in cattle. In 1978, researchers identified the association between scrotal circumference in bulls and age of puberty in their daughters (Brinks et al., 1978).

Since then, scrotal circumference has been used as an indicator trait for puberty. Breed association web sites show substantial increases in scrotal circumference occurring from 1985 to the present, indicating great progress has been made through selection for this trait; a similar response in age of puberty would be expected. Indeed, the inability of heifers to attain puberty prior to breeding may not be as problematic as heifers reaching puberty before weaning.

The association between timing of puberty and subsequent pregnancy rate also seems to have changed over time. Early research indicated heifers should experience two or three estrous cycles before the onset of the breeding season because fertility of the first estrus is lower than subsequent estrous cycles. Thus, delayed onset of puberty was expected to be associated with lower pregnancy rates. However, several studies have not shown strong associations between nutritionally related changes in age of puberty and final pregnancy rates.

Evidence for a genetic basis for these differences is provided by Freetly and Cundiff (1997), who reported pregnancy rates were greater in heifers AI sired by bulls born after 1988 than bulls born between 1982 and 1984, but age and weight at puberty were not. These changes, combined with the continued increase in cost of harvested feedstuffs indicate the need for alternative development systems, which allow heifers the opportunity to conceive early as yearlings at reduced cost.

Current Research

Feeding replacement heifers to traditional target weights increased development costs relative to more extensive heifer development systems where heifers were developed to lower target weights ranging from 51 to 57 per cent.

Feeding to pre-breeding weights as low as 51 per cent of mature weight was shown to be more cost effective than development to 57 per cent of mature weight, even though lighter heifers were allowed a 15 d longer (45 vs. 60 d) breeding season (Martin et al., 2007). Extending the breeding season by 15 d for lighter heifers resulted in similar conception rates between systems, but pregnancy rates for the first 45 days of the breeding season, were 89.8 and 77.9 per cent for heifers fed to 57 per cent or 51 per cent of mature weight, respectively.

Further characterisation of non-pregnant heifers within each system revealed 78.9 per cent of open heifers developed to 51 per cent of mature weight but only 45 per cent of open heifers developed to 57 per cent of mature weight were pre-pubertal prior to start of the breeding season.

This lends support to the hypothesis that one of the major determinants to a heifer's ability to conceive during her first breeding season is the age she reaches puberty, especially in relation to the start of the breeding season. Heifers calving early during their first calving season have greater lifetime calf production than those calving late and are more likely to become pregnant sooner at two years of age (Lesmeister et al., 1973).

However, there was no difference in second-calf conception rates between cows developed to 51 or 57 per cent of mature weight prior to breeding as yearlings (Martin et al., 2007). This indicates lighter heifers that became pregnant during the 15 d extension during the first breeding season rebred with similar efficiency as those pregnant within the initial 45 days. Therefore, proportion of heifers retained as pregnant 2-yr olds was similar between systems. Thus, heifers may be developed to lighter than traditional target weights without negative effects on profitability or future productivity.

Research at Fort Keogh evaluating lifetime productivity of heifers developed with either unlimited or restricted (27 per cent less feed) feed during the postweaning period supports the potential to reduce target weights and costs during heifer development.

The association of age at onset of breeding and cumulative pregnancy rate was similar for heifers developed on the two protocols. However, restricted heifers were lighter at a given cumulative pregnancy rate. Thus, age at the beginning of the breeding season was more critical than body weight. Furthermore, rate of growth from birth to weaning accounted for more variation in puberty and AI pregnancy rate than did ADG during the postweaning period. Neither age nor ADG prior to postweaning period influenced final pregnancy rate. Thus, age and early growth rate (up to ~ 8 mo. of age) influenced time of puberty and conception, but did not alter overall pregnancy rate in a 48 to 60 d breeding season.

When summarised over the last 7 years, heifer pregnancy rate was 3.5 per cent less in heifers developed under restricted feeding at Fort Keogh. Restricted feeding during the 140-d postweaning period reduced harvested feed inputs by 22 per cent and increased efficiency of gain. After restriction, restricted heifers remained lighter but had greater ADG. Restricted feeding improved biological and economical efficiency during and after the feeding period.

Pregnant heifers resulting from the two postweaning treatments were also fed at different levels throughout each subsequent winter. Heifers developed without restriction were provided adequate levels of harvested feed from early December through calving while heifers developed on restricted feeding were fed 20 to 45 per cent less harvested feed. Restriction resulted in lower bodyweights throughout 5 yr of age (Roberts et al., 2009a) which may equate to lower maintenance requirements.

Heifer offspring from the two management groups were randomly assigned to restricted or non-restricted protocols resulting in 4 treatments: restricted cows from restricted dams, restricted cows from control dams, control cows from restricted dams and control cows from control dams. Interestingly, cows from restricted dams were 35 to 50 lbs heavier than cows from non-restricted dams at 3 to 5 yr of age, due in part to differences in BCS (Roberts et al., 2009a).

Thus, method of developing and maintaining replacement heifers may influence offspring growth and development. Differences in weight and BCS may also impact longevity. Current data indicate that retention to the 5th breeding season was influenced by dam and cow treatments. Retention was lowest for restricted cows from non-restricted dams (39 per cent), intermediate for non-restricted cows from either restricted (50 per cent) or non-restricted (51 per cent) dams, and greatest for restricted cows from restricted dams (66 per cent).

Preliminary evaluation of the performance of the third generation of calves found that calves from restricted cows out of restricted dams were lighter at birth and weaning by 3 and 13 lbs, respectively. Thus, restricted cows from restricted dams may have a lower level of production and greater fleshing ability resulting in greater retention. Current data indicate that the small decrease in calf output may be more than compensated by increased longevity.

Several similarities exist between the heifer development studies conducted at the University of Nebraska (Funston and Deutscher, 2004) and Fort Keogh. Both locations used similar types of cattle (composites with ~½ Red Angus and ½ continental breeding) and the treatments resulted in development to similar target weights at breeding (53 vs. 58 and 55 vs. 58 per cent of expected mature weight).

Growth rates during the development period were similar between locations for the two treatments imposed and both locations observed approximately a 10 per cent reduction in proportion of heifers pubertal at breeding in the lower input groups. Magnitude of savings achieved by lower target weights was also similar (~22-24$/pregnant heifer). In contrast to the Nebraska research, a slight decrease in pregnancy rate (3-5 per cent) has been observed in heifers under restricted feeding at Fort Keogh (Roberts et al., 2009b). Methods used for restricting rate of development differed between Nebraska (lower quality diet) and Fort Keogh (lower quantity fed) which may contribute to differences in pregnancy. These studies indicate an opportunity to improve efficiency and decrease production costs by decreasing amount and/or quality of harvested feeds used for heifer development. The challenge then, is adapting these theories to common productions systems in widely varied environments. In the Midwest, including Nebraska, South Dakota, Iowa, and perhaps southern Minnesota, the opportunity to graze corn stalks is an attractive option to dry lot feeding.

We (Larson et al. 2008, 2009) have conducted experiments grazing heifers on corn stalks or winter range as an alternative to dry lot feeding. In each study, heifers grazed on corn stalks gained approximately 0.5 lb/day less than their more traditional fed counterparts, whether that be winter grass or a dry lot. It is important to note that grazing heifers were only supplemented with the equivalent of 0.30 lb of protein per day and gained between 0.5-1.0 lb/day of ADG during winter grazing.

However, once placed on higher quality spring pasture, the heifers had the ability to gain 2.5-3.0 lb/day prior to and after breeding. Regardless of these compensatory gains, heifers developed grazing corn stalks weighed 5 to 7 per cent less prior to breeding and achieved 55-60 per cent of their mature weight by that point. They were also lighter prior to first calving.

Maximal gain should not be the major goal in heifer development programs. Producers should strive for a sound, functional, low-cost, and pregnant heifer. Previous research (Patterson et al. 1992) supported the concept that a heifer must reach 65 per cent of her mature body weight for maximal reproductive efficiency. More recent data (Larson et al., 2009) provides evidence that a lower body weight is sufficient for attainment of puberty and pregnancy success. However, puberty was delayed by low prebreeding gain, but pregnancy rates were similar (85-90 per cent) between development systems. Previous research (Byerly et al. 1987) indicated heifers that do not reach puberty before breeding may become pregnant later than pubertal heifers.

This is borne out by conception rates to artificial insemination and may partially explain recent results from Larson et al. (2009). Heifers developed in the dry lot had approximately a 10 per cent greater AI pregnancy rate than corn stalk developed heifers even though overall pregnancy rates were similar. As a result, calving date was pushed back on the corn stalk developed heifers, indicating that non-pubertal heifers were inseminated on the second or third cycle during breeding and became pregnant.

All of these data provide evidence that heifers can be successfully developed into productive cows using low quality feedstuffs. Heifer development costs can be reduced by limiting forage quality or quantity without compromising productivity. Regardless of source, low quality feedstuffs exist in every environment and can be fed to beef cattle depending on stage of production. Moving heifer development out of the drylot, in favour of cornstalks or winter range reduced development cost by $45/pregnant heifer. The take home message should be to find a strategy to reduce forage quality to improve profitability.

Bibliography

Adams, C.E. : *Mammalian Egg Transfer*, Boca Raton, FL, CRC Press, 1982.

Adams, Carol J.: *Animals and Women: Feminist Theoretical Explorations.* Durham, NC: Duke University Press, 1995.

Archana Satarkar: *Food Science and Nutrition*, ABD Pub, Delhi, 2008.

Arora, N. : *Manual of Animal Nutrition*, International Book, 2004.

Arseniev, V.A.: *Atlas of Marine Mammals*, Neptune City: T. F. H. Publishing, Inc., 1986.

Arti Sharma: *Fishes : Aid to Collection, Preservation and Identification*, Daya, Delhi, 2006.

Aruna T. Kumar : *Handbook of Animal Husbandry*, Indian Council of Agricultural Research, 2008.

Baker, Steve: *The Postmodern Animal.* London: Reaktion Books, 2000.

Balram Pani: *Textbook of Animal Chemistry*, I K International, Delhi, 2007.

Basavaraj S. Benni: *Dairy Co-operative Management and Practice*, Rawat, Delhi, 2005.

Baudrillard: *The Animals: Territory and Metamorphoses. Simulacra and Simulation.* Ann Arbor: University of Michigan Press, 1994.

Bekoff, Marc: *Strolling with Our Kin: Speaking For and Respecting Voiceless Animals.* New York: Lantern Books, 2000.

Betteridge, K.J. : *Embryo Transfer in Farm Animals*, Ottawa, Agriculture Canada, 1977.

Billingham, R.E., and W.K. Silvers. : *Transplantation of Tissues and Cells*, Wistar Inst. Press, Philadelphia, 1961.

Bourbon, Richard M.: *Understanding Animal Breeding*, Prentice-Hall, 2000.

Bower, B.: *Fossils may Clarify Mammal Evolution*, Science News, 1984.

Brock, J. : *A Natural History of Domesticated Animals*, Cambridge Univ. Pr., New York, 1999.

Bronson, F. H.: *Mammalian Reproductive Biology*, Univ. Chicago Pr., Chicago, 1990.

Brown, L.: *Cruelty to Animals: The Moral Debt*. London: MacMillan, 1988.

Brown, R. E.: *Social Odours in Animals Reproduction*, Clarendon Press, Oxford, 1985.

Bushnell, R.B. . : *Dry Cow Feeding and Management*, A Western Regional Extension Publication, 1979.

Carroll, R. L.: *Vertebrate Paleontology and Evolution*. W. H. Freeman and Co., New York, 1988.

Clark, Stephen: *The Moral Status of Animals*. Oxford: Oxford University Press, 1977.

Clutton Brock Juliet : *Horse power: a history of the horse and donkey in human societies*, National history Museum publications, London 1992.

Clymer, R. : *Nature's Healing Agents,* PA, U.S.A: Dorrance Co., 1963.

Crawford A. : *Experiments and Observations on Animal Heat,* London: Printed for J. Johnson; 1788.

Daniel, J.C. Jr. : *Methods in Mammalian Reproduction,* Orlando, FL, Academic Press, 1978.

Davis, A. : *Let's Eat Right to Keep Fit,* New York, U.S.A: Harcourt Brace Jovanovich, Inc., 1970.

Degen, A. A. : *Ecophysiology of Small Desert Mammals,* Springer, New York, 1997.

DeGrazia, David: *Animals Rights: A Very Short Introduction.* Oxford: Oxford University Press, 2002.

Devender Pratap Singh : *A Handbook of Beekeeping,* Agrobios, 2006.

Devyani Khemka: *Animal Physiology,* Dominant, Delhi, 2003.

Eisenberg, John E.: *The Animal Radiations,* The University of Chicago Press, 1981.

Ensminger, M.E. : *Dairy Cattle Science,* The Interstate Printers & Publishers, Inc., Danville, 1980.

Escobar, Roberto Calle: *Animal Breeding and Production of Camelids,* Lima, Peru, 1984.

Flowerdew, J. R. : *Animals: Their Reproductive Biology and Population Ecology,* Cambridge Univ. Pr., New York, 1987.

Gay, W.I. : *Methods of Animal Experimentation,* Academic Press, New York, 1965.

Godthelp, *Animal: Riversleigh. The Story of Animals Reproduction in Ancient Rainforests,* Reed Books, Balgowhah, 1991.

Goel, A K : *Basic Concept of Animal Chemistry,* Pearl Books, Delhi, 2008.

Gordon, G. A. : *Animals Physiology,* Harper and Row, New York, 1989.

Gray, J.: *Animal Locomotion,* Norton, New York. 1968.

Greene, H. W.: *Mode of Reproduction in Lizards and Snakes of the Gomez Farias Region,* Tamaulipas, Mexico. Copeia, 1970.

Griffin, D.R.: *Animal Minds,* University of Chicago Press. Chicago, 1992.

Grzimek, B.: *Grzimek's Animal Life Encyclopedia,* McGraw Hill, New York, 1989.

Hacker, J.B. : *Nutritional Limits to Animal Production from Pasture,* Farnham Royal: CAB, 1981.

Hagedorn, A.L.: *Animal Breeding,* Crosby Lockwood, 1950.

Harrison, R. J.: *Functional Anatomy of Marine Animals,* New York: Academic Press, 1974.

Hulbert AJ, Else PL. : *Mechanisms Underlying the Cost of Living in Animals,* Annu Rev Physiol. 2000.

Hutt, Frederick B.: *Genetics for Dog Breeders,* Freeman & Company, 1979.

Joysey, K. A. : *Development, Function and Evolution of Animal Teeth,* Academic Pr., New York, 1978.

Krieger, Maggie and Richard: *Secrets of the Andean Alpaca - The Field Guide,* Saltspring Island Llamas and Alpacas, 1994.

Lance, J.W. : *Migraine and Other Headaches,* New York, U.S.A: Scribner, 1986.

Lata Bhattacharya: *Animal Biochemistry,* Discovery, Delhi, 2010.

Lyster, S. : *Animals and Their Moral Standing.* London : Routledge, 1997.

Marshall, R.B. : *Breeding Farm Animals,* Asiatic Pub, Delhi, 2006.

Martin, A.M. : *Fisheries Processing : Biotechnological Applications,* Chapman and Hall, Delhi, 2009.

Mathialagan, P : *Textbook of Animal Husbandry and Livestock Extension,* International Book Distributing Co, Delhi, 2005.

Matthews, L. H.: *The Life of Animals Reproduction,* London, Weidenfield and Nicholson, 1969.

Mindell, E. : *Mindell's Vitamin Bible,* New York, U.S.A: Warner Books, 1980.

Montgomery, G. G.: *The Early Placental Mammal Radiation Using Bayesian Phylogenetics,* Science, December 2001.

Muybridge, E. : *Muybridge's Complete Human and Animal Locomotion,* Dover Publ., New York, 1979.

Nyholt, D.H. : *The Vitamin & Herb Guide,* Alberta, Canada: Global Health Ltd., 1992.

Rathnakumar, K : *Fish Processing Technology and Product Development*, Narendra Pub, Delhi, 2008.

Raymond, F., Redman, P., & Waltham, R. : *Forage conservation and Feeding.* Ipswich: Farming Press, 1986.

Renaville, R and A Burny : *Biotechnology in Animal Husbandry*, Springer Pub, 2008.

Rhykerd Charles L. : *The Cycles of Plant and Animal Nutrition*, Scientific American Books, San Francisco 1976.

Robinson, Roy: *Genetics for Dog Breeders,* Pergamon Press, 1990.

Safley, Michael: *Synthesis of a Miracle*, Northwest Alpacas, 2002.

Schiller, A. L. : *Anatomy of the Guinea Pig*, Harvard Univ. Pr., Cambridge, 1975.

Seidel, S.M.: *New Technologies in Animal Breeding*, Orlando, FL, Academic Press, 1981.

Shagufta Jamal and H P S Arya : *Participatory Rural Appraisal in Agriculture and Animal Husbandry : A Training Manual*, Concept, 2004.

Short, R. V.: *Reproduction in Mammals*, Cambridge, Cambridge University Press, 1972.

Shukla, M K : *Brain Teasers : Multiple Choice Questions on Animal Husbandry and Veterinary Sciences*, International Book Dist, Delhi, 2007.

Singh,. G : *Chemistry of Amino-Acids and Proteins*, Discovery, Delhi, 2007.

Stuart Patton : *Principles of Dairy Chemistry*, Huntington, N.Y.: Krieger, 1976.

Thornhill, Nancy W.: *The Natural History of Inbreeding and Outbreeding*, Chicago Press, 1993.

Verma, S.R. : *Nature: Fish Genetics and Biodiversity Conservation*, Conservators, Delhi, 1998.

White, M.J.D.: *Animal Cytology and Evolution*, Cambridge, Cambridge Univ. Press, 1954.

Yablokov, A.: *Variability of Mammals: Moscow*, USSR, Nauka Publishers, 1966.

Yadav, Manju: *Mammalian Development*, Discovery Publishing House, Delhi, 2008.

Index